WRIGHT,P.A.
Old farm i

OLD FARM IMPLEMENTS

OLD
FARM IMPLEMENTS

BY

PHILIP A. WRIGHT

WITH 40 PHOTOGRAPHS
AND 39 DRAWINGS

DAVID & CHARLES
NEWTON ABBOT   LONDON   NORTH POMFRET (VT) VANCOUVER

ISBN 0 7153 6801 X

Library of Congress Catalog Card Number 73-168335

©Philip A. Wright 1961, 1974

Printed in Great Britain
by Redwood Burn Ltd Trowbridge and Esher
for David & Charles (Holdings) Limited
South Devon House Newton Abbot Devon

Published in the United States of America
by David & Charles Inc North Pomfret
Vermont 05053

Published in Canada by Douglas David & Charles Limited
3645 McKechnie Drive
West Vancouver BC

Dedicated to
OUR FOREFATHERS

CONTENTS

THE PLATES

ix

DRAWINGS IN THE TEXT

ACKNOWLEDGEMENTS

THE author wishes to thank Miss J. Capp and Mrs R. Gulden for typing the manuscript. Except where otherwise stated in the text the photographs are either by Mr Barry Finch, *Sport and General*, or the author.

With the exception of six old engravings, the line drawings are by the author.

PHILIP A. WRIGHT

The Vicarage
Roxwell, Essex

INTRODUCTION

ONE of my happiest annual tasks for the past thirteen years has been to act as host on behalf of Essex Agricultural Society to those who qualify for awards at the Essex Show after having worked forty, fifty and sixty odd years either on the same farm or with the same Master. It is a privilege to talk to them and to hear about their early days. Luckily I worked many years on a farm and at fifty-two can say that even in my lifetime I have seen the greatest possible changes brought about by the rapid mechanization of agriculture; and some tools and implements which I used personally are now classed as museum pieces. Guided by my late father I began at a very early age to collect any interesting bygones which came my way and in process of time have built up a folk-museum in miniature. Many of these exhibits form the illustrations of this book—duplicates I have passed on to Reading and to the Science Museum. It is difficult to know quite where to draw the line—ancient and modern overlap in a remarkable manner. A smallholder today may be broadcasting some seed in a field next to which the most modern combine-drill is at work; and a mediaeval barn may be used for storing the very last word in combine harvesters.

Excavation of Roman and Early British villages has also proved that if a man who had lived 2,000 years ago could walk through our fields today he would find very few tools and implements whose purposes would be completely unknown to himself. Those early folk in fact were quite familiar with scythe, sickle, spade and plough.

The Chaldeans kept great flocks and herds, they let in the waters of Euphrates by means of sluices and flood-gates to deposit top-dressings of silt on their barren land. The Assyrians diligently raised water mechanically from rivers and by ducts and embankments did much irrigation work. The Persians and Chinese did likewise. From the Old Testament we learn that the Phoenicians held the plain of Sharon, the "Excellency of Carmel" and the "Glory of Lebanon", and in selling their superfluous produce became a great commercial community. That very early writer Pliny tells us a good deal about Egyptian farming, and when Pharaoh gave permission (through Joseph) to the Israelites to settle there he added "and if thou knowest any men of activity among them make them rulers over my cattle". Both Old and New Testaments show quite plainly that the Israelites were a nation of farmers. The Classical Greeks gave agriculture a high place and Ceres was worshipped as the goddess of corn harvests. They had a variety of implements which included carts as well as hand tools. An agricultural literature was possessed by the Romans and a Carthagian General named Mago wrote twenty-eight volumes and was ordered by the Roman Senate to translate this huge work into Latin. Cato wrote on rural affairs, as did Varro and Virgil. "In those days," wrote the elder Pliny, "when a man was meant to be highly spoken of he was called a good husbandman and whoever was thus praised was thought to be highly honoured." Roman farm implements were more numerous and better in detail than those of the older nations. Most of their ploughs for instance were complex and had a beam, a cross-bar, a handle, a share, mouldboard and coulter, whilst some even had a wheel. We shall see that they even had a reaping machine very similar to that "invented" by the Rev. Patrick Bell hundreds of years later. Our research will be mostly limited to Britain where our aborigines lived mostly on roots, berries, milk and flesh. The influence of the advanced Romans was soon felt so that by the time of the Anglo-Saxon invasion, Britain was richer in flocks and herds. Subsequently the monks became the best farmers. The Venerable Bede describes an abbot for instance, "who being a strong man and of an humble disposition used to assist his monks in their rural labours, sometimes guiding the plough by its handle, sometimes winnowing corn, and sometimes forging implements of husbandry with a hammer upon an anvil". The Normans on the whole improved our farming, but it was left to the good work of the monks to reclaim vast tracts of forest and fen. By the sixteenth century Sir Anthony Fitzherbert had written *The Book of Husbandry and the Book of Surveying and Implements*. When the first Elizabeth ascended the throne feudalism was dying and big tracts of rich hunting ground were thrown open to the plough. To her reign we owe the magnificent doggerel poem by Thomas Tusser, *The Five Hundred Points of Husbandry*, full of good sense and mentioning carrots, turnips and cabbages which had been recently introduced. In the early eighteenth century came Jethro Tull, inventor of drill husbandry, but like so many pioneers he encountered derision from those he sought to help—we shall speak of him in a subsequent chapter. Ultimately the travels and writings of Arthur Young gave great impetus to scientific agriculture in this country. Contemporary with Young was the famous Parson Woodforde whose diary gives a fascinating picture of rural life between 1740 and 1803. Eighteen years after this William Cobbett began his *Rural Rides*, which gives us an insight into the contemporary country life of Britain.

By grouping the following chapters under the main headings of the four seasons I shall try to show something of the methods of an age which has passed from us with startling rapidity. Men are always inventing new methods of saving labour and increasing speed. I have been amused and amazed to find however that some "new" inventions were known hundreds of years ago in part and then apparently shelved for generations. By going back to the land for instruction we are soon reminded that if the land is starved by greed and exploitation, there follows disease of soil, plant life, animals, poultry and man himself. We must ever seek to advocate the virtues of good husbandry and the love of man for his native soil—after all we inherit it from the dead, and we hold it in trust for the yet unborn. The true countryman is familiar with the varied complete process of nature, as well as with the phenomenon of well-founded expectations failing to mature—in a word he is in touch with reality. He is master of a dozen inherited crafts; many of the survivals which I shall mention in subsequent chapters are examples of individual craftsmanship which stand out in pleasing contrast against modern mass-produced implements.

SPRING

SPRING

CHAPTER ONE

I often try to imagine the scene when prehistoric man first made the profound discovery that he had the power to plant a seed of something he really wished to grow and could step back and watch the fruition of this initial effort. No longer was nature in supreme command, but this of course was at a period unknown to historians. From the first stages of our known history a "sower went forth to sow", and for many generations he used his Biblical basket slung around his neck and scattered the seed broadcast.

FIG. 1. Seed-basket

In my lifetime I have watched this time-honoured operation, the only difference being that a wooden seed-lip or sometimes a tin one was used instead of the traditional basket; and the same receptacle served for artificial manures. This was never a haphazard method, but the measured step, the regular manipulation of the hand and the casting of the seed was learned by long observation and practice. Experienced men could thus regulate the prescribed amount of seed to the acre with admirable precision. The real expert, in fact, sowed with both hands, throwing the seed both right and left. A later development was the seed-fiddle which consisted of a bag in a wooden frame and a revolving disc rotated

3

by a thonged-bow. I had one of these fiddles and passed it on to the Science Museum, and to this day a firm in Scotland still makes them for sale. I did however retain my broadcaster, which is similar in design except that a beautifully made system of cog-wheels and bevel gearing rotates the blade at the turn of a handle.

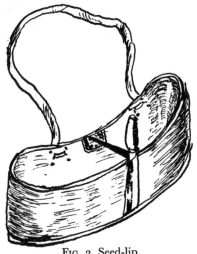

FIG. 2. Seed-lip

It is the only one of its kind I have ever seen and was given me by Mr. Wilfred Pulham of Little Bealings. I have spoken to elderly farm-workers in my boyhood who had actually used a dibber, and this implement was hailed as a great discovery in its native Norfolk. It appears that a tenant farmer on the Hethel Hall Estate in that county had dibbled some holes in his garden and planted some wheat grains which until this time had always been scattered broadcast. It rather pleased him to note the resultant crop becoming superior to the scattered plot. Next year this pioneer got his local blacksmith to make him a dibbling-iron (often called a setting-iron) rather like a miniature crowbar. He set half an acre in this manner, a few neighbours followed suit and the rest (as always) laughed him to scorn. Better crops and more economical sowing gradually led to its adoption, and before long a Norfolk news-sheet got hold of the story and men on the nearby Holkham Estate of the famous Coke became interested. The use of the dibber on freshly ploughed clover land was most successful.

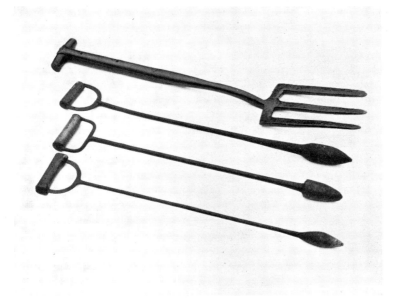

SUBSOIL FORK AND THREE DIBBERS

SHOES FOR INJURED FEET, PALFREY AND OXEN SHOES

BEAN BARROW

BROADCASTER, MADE IN 1780

Sometimes the little implement was called a dibbler, but usually this term described the man doing the job.

There was a recognized method—a man held an iron in each hand and walked backwards making holes about four inches apart each way and an inch deep. Into these holes the "droppers" placed two wheat grains and in this manner big fields were planted. The "droppers" were often a man's wife and family. In fact it represented almost a second "harvest" for him in cash value. In later years a labourer could earn up to five shillings per day out of which he had to pay his "droppers". By no other method could he at that period earn a similar amount so easily on the farm. The job was obviously a tough one however, and very hard on the back. A "bush-harrow" was drawn over the dibbled acres—usually an old gate or hurdle laced with bushes. One dibbler with two or three droppers could set an acre in two days, if the weather conditions were favourable. On the lighter soils in dry weather the job was almost impossible, as the holes filled up as soon as the iron was withdrawn, but then wheat was always reckoned a *heavy*-land crop. When first practised, this method was said to save at least two bushels per acre in seed as compared with broadcasting. In addition the crop gave less "tail" corn and being a better sample would command a higher price. It was estimated that if the method were universally adopted the corn thus saved would provide bread for more than half a million people. The grain thus planted germinated more evenly and was easier to hoe in rows. Old men have told me too how remarkably straight the rows were and how skilful the dibblers became. By the beginning of the nineteenth century few Norfolk and Suffolk farmers practised broadcasting so long as they could get labour for dibbling.

By this time Jethro Tull had long since invented his drill, but as in the case of all pioneers his great contribution to seed-time was hardly noticed in his lifetime. Tull (1674–1741) was born at Basildon, Berkshire, took an Oxford degree and was called to the Bar. Illness drove him back to the country and on Sundays he played the church organ. He got his idea of the drill feed-pipes from the organ pipes—he felt that what went up would also come down—and so his first drill not only planted corn at a suitable depth but covered up behind more or less with an even regularity. The seed-box of the drill was generated from an idea which Tull gained from studying the sound-box of the organ, and linking it with a wheelbarrow.

In the year 1733 Tull published his book "The Horse-Hoeing Husbandry; or an Essay on the Principles of Tillage and Vegetation; wherein is shown a method of introducing a sort of Vineyard Culture into the Corn-Fields in order to increase their product and diminish the Common Expence; by the use of Implements described in Cuts". A formidable title and a weighty book indeed.

FIG. 3. Horse-hoe

An interesting feature of this startling innovation was that Tull appeared to adopt the practice of drilling only as an aid to hoeing by horse-power between the rows. Had field-drilling begun on its own merits with horse-hoeing merely as an accessory it would, I think, have been far more speedily adopted. As it was however, the system suffered far too much ridicule and suspicion and farmers continued with their dibbling for a very long time rather than take note of the new-fangled notions of a lawyer-organist! The principles of Tull's drill are embodied in the tractor-drawn drill of the present day, although many now have steel wheels. The famous firm of Smyth of Peasenhall however still makes a magnificent drill with wooden wheels and in long years has found it necessary to make very few alterations to the original design. One of the "discoveries" of our present century is said to be the combine-drill which places the artificial manure in the trench simultaneously with the seed. Strange to say however "there is no new thing under the sun", for well over one hundred years ago Messrs. Garrett of Leiston were making a combined seed and manure drill which was an excellent machine and did both operations at once. According to an old drilling song only one seed in four is likely to reach maturity: "One for the rook, one for the crow, one to rot and one to grow." The late Harry Gooch used to tell me of old farmers who refused to sow barley until the land was warm, and who tested it by lowering their trousers and sitting down. Mr. Stanley Peck of Colchester tells me his father recalls similar incidents in Essex. We used never to drill beans but ploughed them in with a bean barrow and I show a photograph of one of these now obsolete implements given me by John Foulds of Epping.

FIG. 4. Levelling-box

The earliest form of harrow was probably a stiff spinous plant, such as hawthorn, cut to lie flatly on the ground and make a number of parallel scratches. Very ancient monuments, however, show a wooden frame of bars and crossbars with projecting teeth also of wood. Although the material used eventually became iron and weight and size were changed, yet that early principle has remained with us. The main use of the harrow was always to break down a seed-bed into fine tilth, or to follow the drill to give greater coverage, and later still to kill weeds. When I was a boy, however, it was customary for the meadows to be given a harrowing in the spring, and Charlie Long would interweave a set of harrows with bushes and branches so as to present a rough, dense brushy surface.

All the early rollers were simply cylinders used for breaking down clods of stiff land preparatory to making a seed-bed or rolling down a loose surface after drilling, or to conserve moisture in a dry season. We no longer see the old wooden rollers of my early days. I recall seeing one once which was merely the full diameter of a large tree. These would not make an impression on concrete-like clods, but were ideal for compressing the earth around newly-sown seeds such as turnips and swedes. Some were spiked with iron or encircled with large rings. According to their size and weight, they were drawn by a couple of horses or from four to six oxen. Stone or granite rollers were common in a district where such material was available. Iron rollers varied a good deal and were either flat, spiked or ribbed—the latter were most popular in Suffolk and Norfolk and are still in use there. Before the advent of steam, horse-drawn rollers were the only means of road repair apart from just laying a load of stones for the traffic to smooth down as best it could.

CHAPTER TWO

SHEPHERDING is one of the few farm jobs still not mechanized and one of the oldest callings, which in character has sustained few changes. An Eastern shepherd in patriarchal times was a most interesting character and of course formed the subject of sublime allusion in Biblical and other literature. Today in Britain a hill shepherd is eminent in commonsense and skill; weather lore and sheep lore are his very life. In my schooldays a boy would leave school to become a "shepherd's page" and at lambing time he would lodge in the wheeled shepherd's hut. Shepherding also has a tragic side, for on the road from Newmarket to Bury St. Edmunds there is a grave with a history. It lies at the cross-roads and the generally accepted theory is that it contains the body of a gipsy boy who had the responsibility of keeping sheep. He is said to have accidentally lost one of his charges overnight and in that harsh age the penalty for sheep-stealing was death; so he panicked and hanged himself. To this day gipsies tend his grave with the utmost care.

As a boy I often held the lantern for my father when our tiny flock of Southdowns were lambing; but the native breed of Suffolk was of course the big black-faced Suffolk variety. Compared with the hilly North, East Anglian shepherding has always been different. Crops are specially grown to be eaten-off in sections. Here the hurdle-maker made his contribution and this skill and the tools he used will form the subject of a later chapter. Many of Britain's finest churches are built from the profits from the sheep's back in mediaeval days. On the north doors of my former church at Littlebury, Essex, are carved two pairs of the old scissor-type shears. A few years ago I had the good fortune to rescue a pair of these from a Chelmsford scrap-yard. They are photographed alongside a pair of the later spring-tined type.

It was not until I had acquired a Palestinian shepherd's rod that I appreciated the psalmist's phrase "Thy rod and Thy staff comfort me". I could not see how a rod could offer comfort, especially when we sometimes speak of the "Rod of affliction".

9

The rod in fact is (as will be seen from the photograph) a club suspended by a leathern thong from the waist and used for killing a viper or any other of the sheep's enemies. The staff in Palestine was a straight pole, useful for fording a river as there were no bridges there. In Britain the staff became a crook, a useful implement for securing an individual sheep without disturbing the flock, if used with care. My own crook is of iron, but in Scotland there are beautiful specimens made from horn, and a whole book could be written about their variety. At Littlebury my parish registers went back to 1552 and round about 1680 there were references to persons being "buried in wool", which was made compulsory for a period owing to the poverty which had descended upon the wool trade. A quaint old custom was to put a tuft of wool into the hand of a dead shepherd at burial. This it was said would explain his profession and consequent absence from regular worship when he met his Maker. A delightful story is told of the first Queen Elizabeth on a visit to Sudeley Castle, Gloucestershire. A Cotswold shepherd met her with these words. "This lock of wool, Cotswolds' best fruit and my poor gift, I offer to your highness; in which nothing is to be esteemed but the whiteness, Virginity's colour; nor to be expected but Duty, shepherd's religion."

The mention of sheep is a reminder of the tinkling bells worn around their necks. Many people assume that the very first bells were church bells; this is not so. When man first began to transfer some of his interest from being a huntsman to becoming a herdsman, he needed something to hang on the necks of his animals. This was the circular "crotal" bell, and I have some fine specimens cast by Robert Wells at Aldbourne, Wiltshire, in 1780 and sent me by Mrs. Lott of Swindon. From the sound of the bell comes its name, connected with the verb "belare", signifying the bleating of sheep. These crotals were often used by carters on their horse-harness, whilst sheep-bells developed into the "clucker" or canister types—specimens of both are shown in my photograph, but the most interesting is undoubtedly the larger cow-bell shown. It is riveted and its tongue is made from an animal bone. I have shown it to scores of experts who tell me they have never seen one like it before; it has a very melodious sound. This unique bell came from the late F. J. Hicks of Ixworth.

Fifty years ago horses were the mainstay of farm transport together with draught oxen. Even in the 1930's they held their importance, but pessimists today feel that the farm-horse is doomed to extinction. All honour to the farmers who still breed and use them and to the big brewers and other firms who maintain excellent stables and exhibit them at agricultural shows. I shall refer to horse-drawn implements in detail throughout the book, but I refuse to look upon cart-horses as relics of a past age, and it is a joy to serve on the committee of the London Cart-horse Parade which is one of London's great spectacles at Regents Park every Whit Monday. Here may be seen the cream of the heavy breeds in their glory.

FIG. 5. Bird-scaring Clappers

Bullock or ox transport is extinct in England, but I treasure some nice oxen-shoes or queues. For road work a blacksmith would shoe an ox as he would a horse. The oxen-shoe consists of two plates, one for each part of the divided hoof, as will be seen from the photograph. These specimens came from Mr. T. Okey of Titsey, Surrey. Biblical references speak of the ox as drawing the plough and treading out the corn; all through the centuries oxen were recognized in this capacity. Although slow movers they were steady pullers, extremely strong and not easily frightened or upset. Their harness was a finely shaped yoke to which was attached a long pole fixed to the cart or plough. When horse-breeding improved, oxen became less popular.

Fifty years ago the first job usually given to a boy on the farm was that of bird scaring. He often combined it with the work of "backhouse boy" at the farmhouse, chopping sticks and bringing in coal, wood and water. For bird scaring he would spend long hours with a pair of clappers and earn perhaps 4d a week. My set of clappers was given me by Mr. Harry Cranwell of Chrishall, and consists of an oblong bat to which is bound a square of wood on either side tied together through holes by a leather thong. I have been amused at the so-called "newly invented" forms of bird scarer which have been a source of annoyance to many country sleepers in the past six years by the constant banging right through the night. In my very rare record of 1857 a similar contrivance is described under the title of a "rook-battery". Before that time attempts had been made by steeping old dry rags in a solution of gunpowder and placing them on the windward side of a field, but

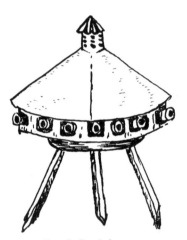

FIG. 6. Rook-battery

the renewal was troublesome. The rook-battery, however, consisted of a circular plate of strong tin 18 inches in diameter, upon the circumference of which was soldered a strong hoop of equally strong tin. This was 3 inches high and through it were pierced twenty-four embrasures each being ¾ inch square and equidistant. At each embrasure a small brass cannon was mounted to a platform soldered to the bottom plate. The cannon were 4 inches long and a conical tin top covered the plate and rim. Surmounting this top cover was a cylindrical lantern. The cannon were loaded with gunpowder and wadded down. Firing was by means of cotton thread dipped into a solution of saltpetre, the thread being held on to the touch-hole of each cannon by copper-wire attached to

SHEPHERD'S ROD, SMOCK AND STAFF

PALESTINIAN PLOUGH AND HOE

A PAIR OF SCISSORS SHEARS, CIRCA 1760 (LEFT),
COMPARED WITH A MORE RECENT TYPE

the platform. Between the firing of each cannon the match thread became longer or shorter. Perpendicular partitions of tin formed numerous channels along which the match thread had to journey so as to reach the touch-hole at the given hour! This wonderful invention stood in the corn on three legs, and was recommended to be moved daily in case the rooks became too used to its being there. I have tried to sketch this almost unknown contrivance.

In 1870 Messrs. Carrett Marshall & Company of Leeds made an alarm gun to drive off hares and rabbits. The inventor was David Joy who gained fame with his valve motion for traction engines and railway locomotives. His alarm consisted of a set of hammers striking percussion caps of a large size. The hammers were worked by a rod furnished with tappets and actuated by a weight, the speed of travel being regulated by a *cataract* cylinder filled with oil. The inventor and a farmer sat up the first night to see its effect on a rabbit-infested field. When the first explosion occurred, rabbits ran off in all directions; at the second they went a shorter distance, and by 4 a.m. they were sitting around the machine watching its action with the greatest interest.

SUMMER

SUMMER

CHAPTER THREE

In July, 1959, following my custom, I took my annual week's holiday coincident with the Royal Show at Oxford. Taking an address at random from a list supplied by the University's Faculty of Agriculture, I found myself lodging at an historic house indeed. Here at 19 Holywell live Mr. and Mrs. Webster in what was once a farmhouse built 500 years ago, long before the city was built up. It was once occupied by John Pinfold, the farmer-grazier-butcher upon whose pasture the very first Royal Show took place on 16 July, 1839; then the whole Show was contained on seven acres where now stands Mansfield College. On my first evening there, the Princess Royal planted a mulberry bush to commemorate John Pinfold and his "Pasture". I subsequently attended the Show which, in 1959, covered 143 acres. On the stand of Messrs. Ransome, Sims & Jefferies, I was told that their firm was the only firm left who exhibited at that far off first-ever venture in 1839. They then sent six tons of machinery on horse-drawn wagons from Ipswich to Oxford, and they have exhibited at every Royal Show ever since!

All English Agricultural Shows, however, owe their origin to the so called "Sheep-shearings" organized by "Mr. Coke of Holkham", as he is described. It need hardly be said that Coke revolutionized farming by his introduction of the famous Norfolk four-course rotation. These "Sheep-shearings" were far more than their name suggests, for here were exhibited livestock and newly-invented or improved implements. Ploughing contests were included and it is of interest to note that the early advertisements stipulated ploughing with "a pair of oxen" which suggests that horses were not yet universally used. The first of these meetings took place in 1778.

At the same period Francis, Duke of Bedford, was organizing

similar events at Woburn, Bedfordshire. I have been into this subject pretty deeply and I never feel that full credit has been given to a man who, from 1790 to 1821, held the post of Surveyor to the Dukes of Bedford. His name was Robert Salmon and he was a constant exhibitor at Woburn Shearings. In 1797, the Society of Arts awarded him 30 guineas for a chaff-cutter and in a subsequent chapter I shall show how slow farmers were to take it up. Salmon also produced the Bedfordshire drill, a fine plough, a self-raking reaper and a thresher, none of which attracted public notice, and some of which were not introduced until sixty years later.

Although it is not the purpose of this book to mention every Show, nor the subsequent formation of the Smithfield Club which included implements, I must mention my very fine copy of the Catalogue of the 1862 Exhibition (there were several editions of a Catalogue, some poor and devoid of illustrations). Class nine was devoted to agricultural and horticultural machinery, and much of this will be dealt with in the appropriate chapters. It is sufficient to note that British manufacturers were showing the widest selection of implements, and these were often so well made that in my farming years in the 1920's and 1930's I was still using some of them. Messrs. Garrett's famous Suffolk firm at Leiston were advertising in five different languages with an obvious eye on the overseas market.

In 1867, the Royal Show came to Bury St. Edmunds and here were portable and traction engines in great number with threshing machinery. Porter's of Lincoln were showing a rotary spade-digger, similar to something re-introduced recently, Burgess & Key of Newgate Street, London, had a big selection of reapers, and Robert Boby of Bury had a fine corn dressing machine.

Messrs. Smyth of Peasenhall were exhibiting their now-famous drill, and Howard's of Bedford had a combined reaper-mower. With the modern emphasis on drying corn from the combine-harvester it is fascinating to note that Mr. Creasey of Wickham Market was showing a drying machine for damp grain. It was so constructed that a large volume of heated air could be passed through the corn at any convenient temperature, while the corn was being gently agitated. A set of blades was arranged in the

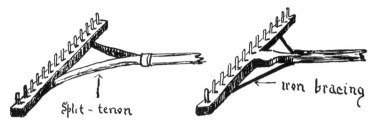

Split - tenon

iron bracing

FIG. 7. Types of rake

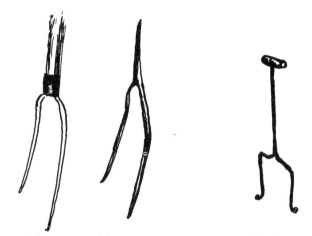

FIG. 8. Two types of fork

FIG. 9. Sheaf-gauge

hand - hoes

FIG. 10. Types of hoes

Weeding - hook.

FIG. 11.

cylinder like a *hummeler* to pass the grain along and to break it up. The outer cylinder, being perforated, allowed the dirt to pass out. All of this sounds so very modern that it has set me wondering why such a machine did not become a huge success. Messrs. Wilkins showed a cast-iron grinding mill said to be capable of grinding 20 bushels an hour. Messrs. Crosskills of Beverley showed a bone-mill and called attention to the fact that some poor man who had had his arm drawn out of his body by one of these mills five years previously was *still alive*. This mill therefore became a prizewinner.

The agricultural shows of today still form a great attraction in the months of June and July and more than ever have added a vast acreage for machinery. This is particularly true of the "Royal" where there is in my view too much duplication in these modern times.

Horse-hoeing and hand-hoeing have been greatly superseded by poisonous spraying for checking weeds. Nevertheless especially with root crops hand-hoeing is necessary and hoes do not seem to have changed in a century and thus will not need mentioning. One little implement to die out, however, has been the weeding-hook which I sketch on page 19. We used to take them into the tall standing corn to cut out thistles which earlier hoeing or horse-hoeing had failed to remove.

Haysel (haytime) was always a busy period in the summer. It was a long and gradual process before the common scythe gave way to the early types of mowing-machine. At Bryers Farm, we always turned hay by hand-forks and rakes (see sketches); I also include a picture of a very pleasing rake made at Pateley Bridge, Yorkshire, by Mr. Harry Longster. I first saw this type used by Walker Brothers of Blayshaw Farm, Lofthouse-in-Nidderdale, and Mr. F. Skaife of Darley has recently sent me the specimen shown in the picture, but there were hay-tedders and swathe-turners on the larger farms and, of course, the coming of the horse-rake helped relieve the drudgery of haytime. Thomas Smith of Bradfield, Suffolk, brought out an ingenious horse-rake just over 100 years ago. The principal feature of his invention was the combining of the teeth or tines with counter-balance weights which tended to raise the teeth from the land, preventing their sinking

ANCIENT SHEARS CARVED ON LITTLEBURY CHURCH NORTH DOOR

BLOWER OR WINNOWING MACHINE

SHEEP-BELLS, COW-BELL AND HORSE-CROTALS

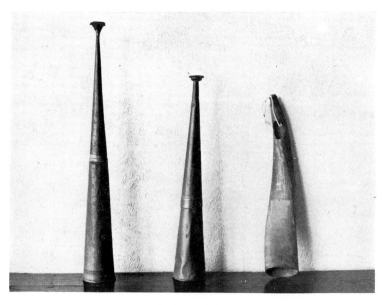

HARVEST HORNS

in. By this method due strength was given to the tines without increasing the pressure on the land.

Records show us that by the twelfth century carts were in use on farms in Britain. The early carts were drawn by oxen by means of the wooden yoke across their shoulders. This yoke was attached to a central pole, and this in turn was fixed solid to the axle. The floor of the cart would likewise be built solidly to the axle and the sides no doubt were woven rushes between uprights driven into the floor. Carts were not used for travel earlier as it was considered bad form, and besides a condemned criminal would thus be taken to his place of execution. The early wheels were of course solid, and cumbersome, as compared with the beautiful craftsmanship of the later country wheelwright. I have tried to show an early type of tumbril which we used in Suffolk for all sorts of farm cartage. The whole of the buck or body was made to tip by with-drawing the pole shown leaning against the wheel which normally held down the extended portions between the big staples affixed to the top of the shafts. These vehicles were of great variety and a whole book could be written upon the subject of carts and wagons; I will merely note the chief differences. The tumbril to which I have already referred would carry anything; stones, manure, farm produce and of course hay and corn. In the latter cases it was necessary for "ladders" to be affixed fore and aft in order that a greater load could be carried. Clumsy though they looked, these carts were of fine craftsmanship and made to balance so well that the load was not thrown too far forward so as to bear down on the horses' back-chain nor on the other hand too far back so as to strain the horses' belly-band or belly "want". The so-called "Scotch" cart was a development and its importance lay in the fact that its tipping power could be gradual. The subject of my sketch was either completely tipped or completely upright. A spring contrivance with a series of holes was erected in front so that the load could be discharged gradually, a little at a time. This was specially important if the load was, for instance, fat pigs. On many a Monday morning I was dispatched as a boy from Hawstead to "Garrard's" the butchers at Hartest with a load of pigs, and they were big porkers too in those days. Covered with a pig-net we would jog along behind Snip—a beautiful strawberry-roan mare.

FIG. 12. A Suffolk tumbril

FIG. 13. A typical wagon

The four-wheeled wagon gave even more scope to the village wheelwright. Here again there was variety, craftsmanship and beauty. Farm wagons were very diverse according to the custom of the counties or to the job they were meant to do. A typical Suffolk wagon was somewhat heavily built as may be seen in my sketch. Rough measurements would be 12 feet long (without the extending ladders), 4 feet wide and about 2½ feet deep. The heavier type would cart wheat in sacks to miller or merchant, and would take its place with the lighter made harvest wagon at haysel or harvest-time. When a farmer or farm-worker removed, the wagon became his furniture van, and often, painted up and festooned with flowers and black crêpe, it would convey his coffin to its last resting-place in the village churchyard.

The Sussex type wagon stands about 6 feet high at the back, 5½ feet at the front and measures nearly 14 feet in length. Its floor is often narrowed behind the front wheels to allow the wheels to lock round in a short curve; this of course lessens the load capacity of the wagon. In the Suffolk type the wheels lock into specially made recesses seen in my sketch. The Hampshire wagon is similar in type to the Sussex. The Woodstock or Oxfordshire type has a convex curvature of the rails over the hind wheels which gives it a symmetrical appearance, whilst the Berkshire type was similar but smaller. A narrow wheeled hooped wagon was peculiar to Wiltshire and Gloucester with very high rear wheels.

The work of the wheelwright began in the woods where he would buy and fell the trees. These were carted home by a horsedrawn timber drag and put over a saw-pit. Here the planks would be cut—until the advent of the steam saw—laboriously by hand. I have a crosscut saw which was thus used at Audley End timber yard; strictly these saws were not "crosscut" as this type only cut on the downward pull.

All the earlier cartwheels and wagon wheels were of wood, and so too were the first axles. A brief description of the making of such wheels might not be inappropriate. I have watched the process outside the blacksmith's and wheelwright's at Whepstead as a boy. A typical wheel consists first of the "nave", which is the name given to a turned block of wood (elm or oak) bound at each end with iron hoops and pierced with twelve mortises equally spaced

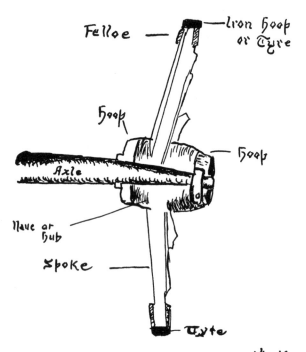

Felloe — Iron Hoop or Tyre

Hoop

Hoop

Axle

Nave or Hub

Spoke

Tyre

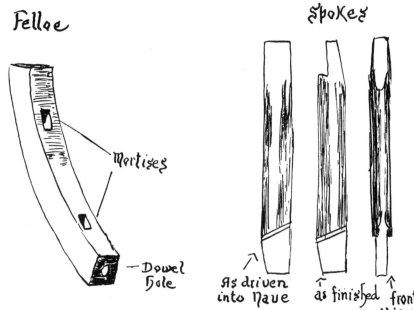

Felloe

Mortises

Dowel Hole

Spokes

As driven into Nave

as finished

front view

FIG. 14. Parts of the wheel

around it. Into these are inserted a similar number of spokes, these always being made from best seasoned oak. A ring of wood in six segments or "felloes" is then mortised upon the extreme ends of the spokes, and this is finally surrounded or "shod" by an iron hoop or tyre which can either be in one stout unbroken ring or in segments. The hoop or tyre is always applied hot, and its contraction when cooled brings the felloes into close contact with themselves and the iron and thus of course adds considerably to the strength and durability of such wheels. It will be noticed that the face of the wheel was designed concave or "dished" and there was good reason for such a plan. A cart or wagon had to encounter rough, unmade roads, potholes and rough land. Such constant bumping could mean the whole of the considerable weight of the load being thrown at one thrust on the inside of the stock (sometimes called the nave or hub). If the wheel were quite flat, such sudden jolting would easily knock out the spokes, so the wheel was dished. The stocks were chipped out by hand before the days of the lathe, and the axe and adze would be used. The spokes would be cleft and not sawn and were tenoned into the stock, one forward and one backward so as to "spread" the load. As these spokes were made to radiate and had to lean out a bit, it was a matter of sheer skill to cut out the mortises. The rims or felloes could be either of elm, beechwood or ash and, as we have said, these had to be made in sections, dowelled together and mortised to take the spokes. Because of the "dishing" of the wheels the spoke at the bottom would be vertical and the other one would be leaning out slightly. The felloes had to be thus fitted so as to pick up the spoke tenons and at the same time to make a circular rim to receive the tyre. To measure the circumference of the wheel and tyre, the wheelwright used a "traveller", the name given to a very ancient tool. It consisted of a wooden wheel 10 inches across with a handle attached to its axle; by running the wheel round and making sundry chalk-marks and calculations an accurate measurement was made. Even the latter had to be slightly conoid in shape. I hope my sketches will make this last paragraph clear. I do not apologise for making this slightly technical aside, for the following reasons. Almost every cart or wagon and many horse-drawn implements, including seed-drills and horse-hoes, had this type of wheel. Even the farmer's light cart or pony trap in which he drove to market would have a similar wheel, although smaller, lighter and neatly made. Messrs. Symthe, the famous Suffolk drill manufacturers, still turn out their drills with such wheels today, but it must be admitted that even horse-drawn carts and wagons today are more and more made to run on pneumatic tyres. I show a picture of a cart-jack for use when greasing cart-wheels; it came from Percy Wedd of Audley End. Perhaps a mention should be made at this stage of the close colleague of the village wheelwright —the blacksmith. His work is very varied today and he is no less than an agricultural engineer. At the same time what horses remain to us are still shod in the time-honoured manner. Fewer by thousands are these craftsmen today and many of them can echo the epitaph on a tombstone at Darsham, Suffolk, in memory of Jeremiah Easthaugh who died on 19 March, 1788, aged sixty-nine and whose memorial reads

> My sledge and hammer are reclined
> My bellows too have lost their wind
> My fire extinct my forge decayed
> And in the dust my vice is laid
> My coals are spent, my iron gone,
> My nails are drove—my Work is done!

CHAPTER FOUR

From the very earliest times wherever corn was grown some sort of a grinding-mill would be found nearby. Primitive people merely pounded the grain between two stones in order to render it palatable. The very first and natural advance was a simple quern or hand-mill. An immovable nether stone was surmounted by an upper stone which was put into motion by the hand. Such machines were in common use among the Hebrews and Greeks and in the latter mythology one Pilumnus Myles or Mylantes was said to have invented the mill. The Romans no doubt brought the quern into Britain and fine specimens of these little mills can be seen in Colchester Museum. Later on watermills were in use by the Romans and windmills had appeared by the time of Augustus. Horse-power for driving millstones eventually became common on farms, to be ultimately supplanted by steam-engines, oil-engine tractors and nowadays electricity. It was, however, far more common for grain to be taken to the local watermill or windmill and if this chapter is largely devoted to the latter variety it is only because I have spent part of a lifetime's study on this fascinating subject, and I have personally sketched or photographed some 150 specimens in this country the majority of which have disappeared today.

A brief description of a stone mill may not come amiss at this stage, because whatever the power used the principle is the same. As in the case of the hand quern the lower stone is fixed while the upper stone revolves with considerable velocity, being supported by an axis passing through the lower stone. The distance between the two is capable of adjustment according to the fineness which it is intended to produce in either the grist meal or flour. When the diameter is, say, five feet, the stone may make about ninety revolutions per minute without the flour becoming too much heated. The corn or grain is shaken out of a hopper by means of projections from the revolving axis, which gives to its lower part or feeder, a vibrating motion. The lower stone is slightly convex and the upper stone somewhat more concave so that the corn,

27

which enters at the middle of the stone, passes outward for a short distance before it begins to be ground. After being reduced to powder it is discharged at the circumference, its escape being favoured by the centrifugal force, and by the convexity of the lower stone. The surface of the stones needs to be constantly "dressed", i.e. cut into grooves in order to make them act more readily and effectually on the corn. This was a skilled job for the miller or the itinerant millwright, who with mallet and hard chisel would work hour after hour chipping and cutting oblique grooves to assist the escape of the meal by throwing it outward. I have often watched this task being done in my boyhood.

The operation of "bolting" by which the flour is separated from the bran or coarser particles is performed by a cylindrical sieve placed in an inclined position and turned by machinery. The fineness of the flour or meal is said to be greatest when the bran has not been too much sub-divided, so that it may be more readily separated by "bolting". This takes place when the grinding has been performed more by the action of the particles upon each other than by the grit of the actual stone.

Domesday book records a good many "mills" and it is generally assumed that the inference here is watermills. The very first certain written record of an English windmill occurs in the famous Chronicle of Jocelyn de Brakelond, a monk of the Abbey of St. Edmundsbury (Bury St. Edmunds), dated 1119. I have always loved this extract and as the Haberdon meadow at Bury St. Edmunds (the scene of the occurrence) has always been known to me, I make no apology for reproducing this extract from Jocelyn's famous diary; the Abbot of course was Samson. "Herbert the Dean erected a windmill upon Haberdon. When the Abbot heard of this, his anger was so kindled that he would scarcely eat or utter a single word. On the morrow, after hearing mass he commanded the Sacrist, that without delay he should send his carpenters thither and overturn it altogether and carefully put by the wooden materials in safe keeping. The Dean, hearing this, came to him saying that he was able in law to do this upon his own frank fee, and that the benefit of the wind ought not to be denied to anyone. . . . The Abbot answered, 'I give you as many thanks as if you had cut off both my feet. By the mouth of God I will not

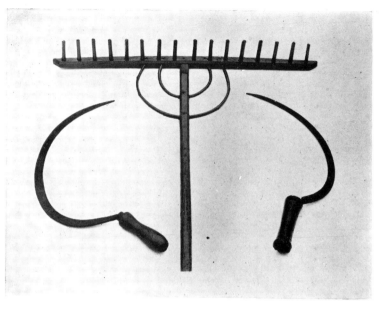

TWO SICKLES AND A NIDDERDALE RAKE

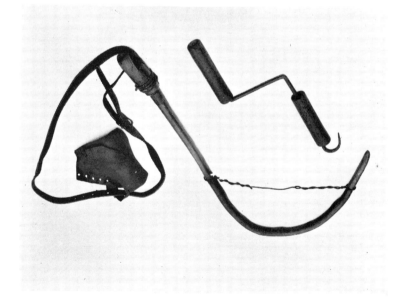

DROWEL AND SCUD-WINDER

EARLY CORN FLAIL, THATCHER'S SHEARING KNIFE

eat bread until that building be plucked down. You are an old man, and you should have known that it is not lawful even for the King or his justicary to alter or appoint a single thing within the banlieue without the permission of the Abbot. . . . 'Begone,' he said; 'before you have come to your house, you shall hear what has befallen your Mill.' But the Dean, being afraid before the face of the Abbot, by the counsel of his son, Master Stephen, fore-stalled the servants of the sacrist, and without delay caused that very mill which had been erected by his own servants to be over-thrown. So that when the Servants of the Sacrist came thither they found nothing to be pulled down." Thomas Carlyle makes the story the subject of his satire in *Past and Present* (1843).

In feudal England the windmills and watermills were important franchises belonging to the lord of the manor, who enjoyed as a rule the sole privilege of "multure", or grinding corn, and levied a toll on all persons using the mill. The miller was the most impor-tant lay-tenant of the manor. The fee for grinding was fixed by the ordinances for trade at a quart of wheat for every bushel ground, which was to be augmented by another bushel if the corn was brought to the mill. Every monastery had its own mill likewise. In mediaeval times the miller was forbidden on pain of fine (and for a third offence the pillory) to either water or change any man's corn or to give worse for better. By the ordinance, the miller's poultry was limited to three hens and a cock and all other grain-gluttons, i.e. geese, ducks and hogs, were often banished from his premises. Latterly things were a bit easier and he was allowed four pounds of meal from every sack or coomb of grist that he ground. This was called the miller's toll. An interesting brass dated 1349 in St. Margaret's Church, King's Lynn, depicts an early windmill, the oldest illustration we have of a primitive post mill. The brass shows a lovely fourteenth-century country scene with a mill to which a customer is approaching with a sack of corn. He is riding on horseback, and carries this sack on his own back to rest the horse, an early example of humour in church illustrating the old saying, "A merciful man is merciful to his beast." There are nice windmill carvings in Bristol Cathedral and in Thornham Church.

As a boy I often stayed at Mill House, Hartest, and day after day I would be in the post mill there with Miller Johnson; the mill then was in regular use. I should describe the three types of windmill—smock, tower and post mills. The smock mill is octagonal and the tower part built of overlapping boards. As with the brick tower mill, only the head which carries the sails turns independently. More attractive are the post mills which predominated in the eastern counties. Here the whole body of the mill

FIG. 15. Smock mill, Upminster

revolves on a centre-post above a fixed roundhouse built usually of brick, sometimes of stone and occasionally of lath and plaster. The earlier type of post mill did not even have a roundhouse, but instead was supported by a tripod of massive beams fixed to even larger beams on the ground. There is a fine mill of this type at Great Chishill on the Essex-Hertfordshire border and this has been occasionally worked by Mr. Richard Duke in recent years.

FIG. 16. Post mill, Syleham

FIG. 17. Tower mill, Pakenham

The working of a windmill (see sketch) is ingenious but very sensible and simple. The sails (called sweeps in some places) rotate a main shaft to which is attached a cog-wheel called the brake-wheel, the cogs being made of hard apple-wood. In turn this brake-wheel engages with a "wallower" which is bevelled. The next wheel below this one is affixed to the centre-post and by means of gears operates the hoist which brings up the sacks of

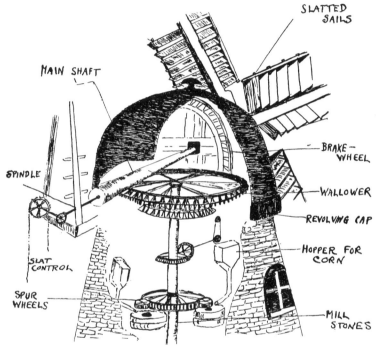

SLATTED SAILS

MAIN SHAFT

SPINDLE

SLAT CONTROL

SPUR WHEELS

BRAKE-WHEEL

WALLOWER

REVOLVING CAP

HOPPER FOR CORN

MILL STONES

FIG. 18. The working of the mill

corn from the ground floor, or straight from the cart or lorry. Below this a big spur wheel engages with the stone-nuts; these rotate the millstones which actually grind the corn. A typical millstone weighs about 15 cwt. and I have already mentioned the need to "dress" these worn stones from time to time. All sorts of cleaning and sifting machinery can be "run off" the system if the mill is big enough.

A mill can only work when the sails are "in the wind". When the wind shifts the sails have to be turned to face it. On the older type post mill the miller did this by climbing into a framework, seizing the extending tail-pole and pushing the mill around bodily. The miller at Drinkstone, Suffolk, demonstrated this for me when that mill was in use over thirty years ago. A later development was the fantail and shafting attached to the ladder which automatically kept a post mill "into wind", and which surmounted the cap of either smock or tower mill for the same purpose.

Ancient mills had merely stretches of canvas on the sails as the sole means of capturing the wind-power. Reefing these sails in a gale was a job equal in risk to that of a sailor on an old time sailing ship's mast. Later on, a clever device of slats or louvres was introduced. These are opened or closed as desired, governed by velocity or lack of wind-power, a weighted chain on ground level being connected with a spindle which goes right through the main-shaft.

Windmills are not the light airy structures they appear to be and they have been responsible for serious injury and sometimes death to both man and beast. I have often heard my father speak of the miller at Lawshall, Suffolk, being drawn into the gearing and killed, and this also happened at Terling, Essex, not many years ago. At Cockfield the mill was blown down during a gale and the miller perished; whilst at Dalham the miller was using his hoist when the mill was struck by lightning and he was killed. Great Chishill mill-sails swept round one day and killed a man and his donkey. The normal weight of one mill-sail is quite often over a ton. It seems incredible to realize that a mill which formerly stood at the entrance to Bury St. Edmunds cemetery was removed bodily to Wickhambrook, and Bocking (Essex) mill was raised several feet by having the body jacked up a few feet at a time and new layers of brickwork added to the roundhouse, giving it greater capacity.

To be in a working windmill is just like being on a ship at sea, every board seems to be alive with movement, and to see a mill at work after dark is most awe-inspiring. I was always sorry that the unique mill at Haverhill, Suffolk, was demolished in 1942. It could be seen from Essex, Suffolk and Cambridgeshire and had

USING A FLAIL

DEMONSTRATION OF AN 1800 BREAST-PLOW

A BINDER

a unique circular or annular sail, the only one left out of the four built in this country. I have a very fine model made from my sketches and photos of this particular mill (see photo). I also have scale models of Herne Bay smock mill and Whelnetham post mill which I have known since I could see but which is now no more. My other model is of the fine tower mill at Pakenham in Suffolk, which, thanks to the hard work and devoted care of the owner, Mr. Bryant, and his sons, is still at work doing the job for which it was built. Mills at Duxford, Dunmow and Great Bardfield are now houses; Stanstead and Hempton are boy-scout headquarters. The mill on Wimbledon Common is one of scores which used to be all over the London district. It is a mixture of all three types, post, tower and smock. Many windmills have been destroyed by fire; stones running empty set up tremendous friction and flour dust is, of course, highly inflammable.

Changed circumstances in this century have, of course, revolutionized the milling trade. Wheat-growing in England prior to the war had been much diminished. The importation of foreign wheat rendered it essential to erect large power mills near the ports of grain-entry. Between 1875 and 1885 an entirely new system of steel rollers for grinding in place of millstones hastened the closure of many wind and watermills. Farmers generally began to have their own mills set up in the barn. Earlier horse-powered mills gave place to steam and then oil-engines. Now this year my own brother's system at Bryers Farm, Hawstead, has changed from the oil engine to a small electric mill. I am, of course, referring to grist for cattle, pig and poultry feeding and not to flour-milling which today is almost always done in a big way in a big mill. Less grist, however, is being home ground today owing to the excellent compounding of rations done by firms like B.O.C.M. who issue their foods in pellet cubes and pencil forms as well as meal.

AUTUMN

AUTUMN

CHAPTER FIVE

FARM work never stands still, and before summer has officially passed into autumn, corn harvest has begun on arable farms. No season has probably been so much romanticized, and writers and poets in every generation have eulogized harvest. What really is the sober truth? In my boyhood it meant aching muscles, frayed tempers, long hours and a complete anxiety neurosis as to whether it would rain again tomorrow. One can therefore hardly imagine the years before my time when there was no mechanization whatever.

Reaping with a sickle was done sometimes by the season, by the day or by the acre and whichever way was chosen it was hard work. The sickles, of which I give a picture, differ from the modern reaphook, in that a true sickle had a serrated edge, weighed about a pound, had a cutting edge of about 16 to 20 inches and a handle about 6 inches in length. It had about 300 serratures all pointing backward toward the handle. One specimen is from the late F. J. Hicks of Ixworth and the other from Mr. H. Law of Wendens Ambo. Sometimes a reaper made his own bands from the cut corn with which he tied his own sheaves. Often reapers worked in pairs and the weaker, either the reaper's wife or some junior member of the family, did the tying.

Mowing with a scythe also required great muscular power and a twisting motion of the body which kept almost every voluntary muscle in constant action. No doubt this job has always been one of the most severe of any farm job. Some of us who have used a scythe for only a fraction of the time spent by these old-timers can testify to its drudgery. Once we had learned to keep the heel of the scythe down, we never really lost the art, and some years ago when a gang of us tackled the churchyard, in my former parish, I overheard one of the workers say, "Well I'm ——— the owd

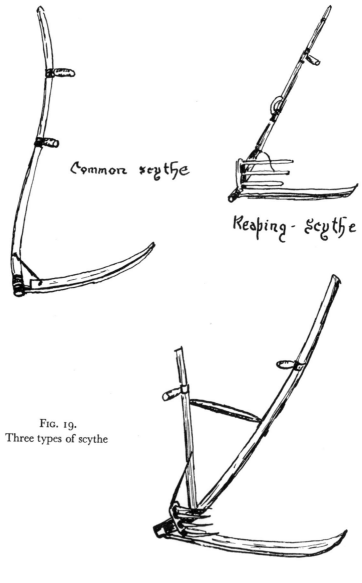

Common scythe

Reaping-scythe

FIG. 19.
Three types of scythe

Cradle Scythe

Parson can swing a scythe", and I blushed at the unexpected compliment.

On an old-time farm the gang of mowers would begin very early in the morning, rest at midday and then continue their mowing well into darkness. The leader of the gang was given the grand old Biblical title "Lord of the Harvest"—the word "Lord" in this case meant to give the lead, and he worked slightly ahead to set the pace. Many a time Charles Long has said to me when we worked together at Bryers Farm, even if the job was only hoeing, "Now then Philip, which of us is going Lord?" Every gang of mowers aimed at a pretty level cut, and their skill was remarkable. The closer in line that all the ears could be left, would facilitate the drying-out of the mown swathe and its position for tying up. I have sketched the three main types of scythe, the common scythe, reaping scythe and cradle-scythe and the sketches are self-explanatory. The cradle assisted in gathering the corn, causing it to fall into a regular swathe. The cradle was fastened down to the snead (handle) with the same iron ring and wedges which already fastened the snead to the cutting blade. The cradle could stand about 13 inches high and have three horizontal teeth or bars of diminishing length from the top to the bottom. The cradle-scythe was even bigger. The gatherers (man, woman or boy) had to do their job quickly in order not to be in the way of the following-on mowers.

It always strikes me as extraordinary that hand-mowing alone was practised until at least the middle of the last century and yet the Romans had a mechanical substitute for hand-labour. From a very old print in my possession, I am able to give a picture of what this machine probably looked like. In the year A.D. 31 Pliny describes this, but Palladius in A.D. 391 gives far greater detail and is worth quoting. "A cart is constructed", he says, "moved on two wheels—the boards in front being lower than the rest, and on that part a great number of teeth are fitted in a row at intervals. An ox is yoked to two poles and harnessed head towards the cart; so that he pushes it into standing corn. The driver walks with the ox and regulates the height and depth of the teeth frame." Palladius concludes his description with these words. "This system is advantageous in open level countries where straw is not much required." All of this seems to me to have a delightfully modern sound and again emphasizes my point that there is no new thing under the sun.

FIG. 20. Roman reaper

We find no mention, however, of any attempt to mechanize reaping in this country until 1780 when Mr. Capel Lloft of Bury St. Edmunds conceived a machine similar to the one described, and was amazed afterwards to discover that the Romans had had such a machine hundreds of years before. In the year 1811, a Mr. Smith of Deanston brought out a machine which excited considerable interest. Between that year and 1837, the machine underwent various improvements but seems to have been ultimately abandoned.

Its cutting principle was an innovation, and consisted of one large continuous circular knife which revolved with great velocity. The gathering apparatus was a revolving rake placed over the cutter and concentric with it. The corn was therefore laid down at one side of the track. Ultimately the machine was made to be pushed by two horses and took a four-foot cut at a time.

By this time, two other pioneers were trying their hand—a Mr. Boyce and a Mr. Plucknet; the former relied upon a number of horizontally adjusted scythes which revolved rapidly; but Mr. Plucknet had a notched sharp-edged circular steel plate which cut the corn like a saw. Neither machine had any proper gathering apparatus and they were soon discarded. Another short-lived machine of the period was Gladstone's, which resembled Pluck-net's, and was made in Castle Douglas. Robert Salmon on the Duke of Bedford's Estate at Woburn was the first to make a machine with the "shears" method of cutting and this gave great promise.

In the year 1822, Mr. Henry Ogle, a schoolmaster of Renning-ton, Northumberland, made a machine in conjunction with a Mr. Brown of Alnwick iron-foundry. It was horse-drawn instead of being pushed and the cutting apparatus projected to the side of the "carriage". By the wheel-motion a straight-edged steel knife was made to vibrate rapidly from left to right. This machine quite obviously was a great advance and yet we hear nothing of its subsequent history. Mr. Joseph Mann of Raby in Cumberland made a reaper in 1820 and ten years later was able to claim that with many modifications there were four "points of Perfection". Draught was taken from the front of the machines; a polygonal knife did the cutting; a series of rakes did the gathering; and the corn was laid out regularly in swathes. In 1832 the Highland Agricultural Society's show was held at Kelso, and this machine was tried out successfully in an oat field, but because the judges did not award a Premium, Joseph Mann became disconsolate.

In 1826, a Scottish minister, the Rev. Patrick Bell of Carmylie in Forfarshire, made a really effective reaper and I reproduce a picture from a book by another parson, the Rev. John Wilson's *Farmers' Dictionary* of 1850. The cutting principle of Bell's machine was almost identical to that used today in binder and combine-harvester—a series of clipping shears. The Highland Society awarded Bell a £50 Premium and in the harvest of 1834 many of his machines were in use, having been made in Dundee. Four of them were sent out to the U.S.A. and became the models upon which many so-called "inventions" of American reapers were based. In 1851 for instance at an exhibition in New York no fewer than six reapers were shown, all by different hands and each claiming to be separately invented. All of them bore the closest possible resemblance to Patrick Bell's machine.

Bell employed one man to drive the horses, and eight women to collect the cut corn into sheaves; four men to bind them and two to set the sheaves up into stooks or shocks as we call them in East

Anglia. It was claimed that 14 acres per day could be cut, but the machine was by no means universally adopted; the report from the Royal Agricultural Society's Show and implement-trials in

FIG. 21. Reaper invented by Rev. Patrick Bell of Forfar in 1827. Five-foot-cut on "clipping" principle by shears; the corn is deposited in neat rows by an endless webbing

1856 was encouraging. The judges said 'there had always been some points of excellence in Bell's machine not shared by any other.'

From this period onwards, many machines came on to the market. Crosskills of Beverley made a three-horse pushing reaper with self-delivery. Much later (in 1872) this firm put out a steam reaper which was pushed by an Aveling and Porter crane traction engine. The knife reciprocated at the base of the platform. I am able to produce a drawing of this unique reaper. Messrs. Dray and

FIG. 22. Aveling engine with Crosskills reaper

Company of Swan Lane, London Bridge, produced Hussey's reaper, and Messrs. Burgess & Key of Newgate Street made one for McCormick. A natural follow-on from the mowing machine was the sail reaper or self-raking reaper. Even today one occasionally hears of one of these being brought out of the cart-shed to tackle a badly laid field of peas. I have a model (see photograph) of the Albion sail reaper.

Subsequently, of course, the self-tying binder was produced and at first encountered bitter opposition from bands of farm-workers who thought it would destroy their livelihood and who wilfully broke up some machines on the farm. Binders are still in use today and it would not be right to describe them here; of course the combine-harvester has displaced them on many farms and in this connection it is of interest to record that forty-six years ago, horse-drawn combine-harvesters were being used in Canada. Canadian and American farmers welcomed mechanization more heartily than we Britishers, but their principles were copied from our own pioneers as I have shown.

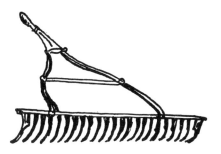

FIG. 23. Hobby rake

The stubble was cleared by horse-drawn carts and wagons (dealt with in Chapter Three on haymaking) and in this connection I must mention my first job in the harvest field. It was to drag the iron-toothed rake which I have sketched and which for some reason was called the "hobby rake" and which seems completely to have become a museum-piece. After the horse-rake had been over the stubble, the womenfolk were allowed to glean the ears of corn. This was not merely for the fowls to eat as in more recent years, but the early wheat-gleanings were taken home, threshed by flail, winnowed and ground for bread in an age when starvation was not unknown. Many parish churches had a custom of ringing the bell to denote that gleaning might begin.

The presence of three interesting "harvest horns" in my collection (see photo) reminds me that this old harvest custom deserves a mention. Two of the horns are of tin and the third is fashioned from a bullock's horn. The longest horn measures 20 inches, the small end being nicely tapered into a mouthpiece. It was sent me by the widow of Charles Copeman of St. James, near Halesworth, who used to blow it at 5 a.m. to call his twenty-two mowers to work in the year 1857. The other tin horn is slightly shorter but very similar in construction. It belonged to the late Mr. William Smith of Great Sampford, Essex. The bullock-horn model belonged to my father, the late Willoughby Wright of Hawstead. He used to explain to me that it was the custom of the "Lord of the Harvest" mentioned earlier in this chapter to sound the horn to call the men to work and to tell them when to stop work. These horns have a fine musical note and would sound from farm to farm, and parish to parish. The custom may quite well be linked with the time when farm-workers lodged and fed at the farmhouse and the horn would call them for meals. I have never heard of the custom except in our corn-growing districts of East Anglia, and I am convinced that it was a survival from very ancient times.

The celebrated Thomas Tusser of Rivenhall, Essex, has already been mentioned in this book. His work was published towards the end of the sixteenth century. In his account of the farmer's occupation for August he says, "blowe horne for sleepers and cheare up thy reapers". I would contend that the custom goes back beyond Tusser and is indeed mentioned in an old English Kalendar attached to "Queen Mary's Psalter" written and illuminated by an English scribe early in the fourteenth century (British Museum Royal Ms 2 B VII).

Here the scene is corn-harvesting for August and three labourers are pictured cutting with their sickles, and the "Lord" or "Bailiff" directs their efforts with an outstretched stick. Suspended by a cord over his left shoulder is his horn. During the process of time, the custom apparently lost significance and in 1889 a Mr. G. C.

MODEL SAIL-REAPER

FOUR MODELS OF WINDMILLS

TWO TYPES OF BARLEY HUMMELERS, CIRCA 1850

Pratt of Norwich complained that the practice of blowing the harvest horn was prevalent among young boys in that district. I once asked Joe Betts of Stanningfield where they bought these horns and if it was a music shop. He replied, "Music shop be blowed, Master, they cost me a tanner each in the old days at the ironmonger's."

I have found no literature upon this subject, but I have recorded years ago that a farmer named Abraham Barnard of Fambridge Farm, White Notley, blew his horn in the eighties to call the men to work, announcing their meal times in the fields in the self-same manner. I have also heard of the harvest horn being blown in the Earls Colne and Maplestead districts in the late 80's. When forwarding me the horn from the late William Smith, his son and daughter explained that he used it regularly at Sampford. A boy doing his first harvest had a nail driven into his boot and was expected to treat his tutors, this was called "shoeing the colt".

As soon as harvesting was complete the men would go around in a gang and press people to contribute a "largesse" which consisted of extra beer money.

Another custom of my boyhood was to place an oak bough on the last load of harvest and this I have often done. In some districts a "corn dolly" was plaited from the last sheaf of harvest and carried prominently to the barn.

Here it was placed at the head of the table at the "horkey" or feast which a good farmer gave to his men after harvest. It was then hung in church and not removed until the next new harvest had produced another one. Here was the continuity link and it all seems connected with the old fertility cults. In a modest sort of way in my present parish (population 8,000) the last "village" in Essex before London, we have revived the custom of a harvest horkey as there are still three fields farmed here at Woodford Bridge by Dick Lewis. My photograph shows Mr. Ernest Mizen of Radwinter with two of the "dollies" which he makes, one of which is now in my possession.

CHAPTER SIX

THE art of threshing is as old as seedtime and harvest, for it was useless to grow corn and not be able to use it. The so-called "threshing-floor" of the Hebrews, Greeks and Romans was not the floor of a barn, but an open space levelled and beaten firm and about 30 to 40 paces in diameter. There are plenty of Biblical references to the operation, probably first done with a cudgel. Later, oxen were used and in obedience to Divine Law the Jews "did not muzzle the ox that trod out the corn" but the ancient Greeks besmeared the mouth of the poor animal with dung to prevent it from eating. Gradually, the flail superseded all other methods, and for many generations was the chief or only threshing implement. I have two nice specimens in my collection (see photograph) and I also show the late John Reynolds knocking out some seed-beans at Bryers Farm in 1937. My flails consist of a hand-staff of ash about 5 feet long; a beater rod or swingle of ash or some other hardwood is attached and measures about 3 feet. Attachment was by a shouldered loop of green ash at the head of the handle, well bound and made to swivel, the actual joint being made by leather thonging. The work required skill and was very hard and exacting. In this country, threshing was done in the barn where there was always a specially prepared floor. Two large doors stood opposite, and were valuable for the winnowing process which followed the threshing, i.e. a straight-through draught was easily created.

Barley with its avils or whiskers was always a problem and needed the chopping effect of a hummeler to separate the corn more cleanly. I have two excellent specimens of these and they are seen in the photograph. One came from Bryers Farm and the other from P. & A. Gunn of Abbess Roding.

I am indebted to Messrs. Ransome for the illustration of a very old hand-power threshing machine (1844) which appears to have been the link between the flail and the powered threshing machine. Ransome's machine got through 10–12 bushels per hour and was commended at the R.A.S.E. Show at Liverpool. Long

FIG. 24. Ransome's hand-thresher

before this time, however, Andrew Meikle of Haddington had invented a threshing machine of considerable efficiency. At first it merely separated the corn from the straw and the whole of the corn, chaff and straw were thrown into an indiscriminate heap. By the addition of shakers and hummelers and fans driven by the same power, an effective winnowing operation was produced. Other threshing machines quickly followed and one of these was mentioned in connection with the Royal Show at Bury St. Edmunds in 1867 and two illustrations are reproduced from one of Tasker's old catalogues.

Tasker's Improved 4-Horse Power Portable Thrashing Machine,
ADMIRABLY ADAPTED FOR THE COLONIES.

FIG. 25 (a)

TASKER'S Improved Horse Power THRASHING MACHINES.

1867. At the Royal Agricultural Society of England's Show, Bury St. Edmund's a Prize of £8 was awarded to W. Tasker & Sons for their Horse Power Thrashing Machine.

FIG. 25 (b)

The threshing drum shown in Fig. 25(a) was purchased by my grandfather and on it my father drove the horses. Only one horse is shown in the engraving, but there are places clearly visible for four horses. The scheme explains itself by the pictures, and the thresher was easily packed up and transported from stack to stack or from farm to farm.

As far back as 1795, a Mr. Wigfull of King's Lynn had constructed a machine which attempted to combine the impulse of the flail with that of revolving beaters; the latter were only loosely attached by short chains and when the drum was in rapid motion they were flung by centrifugal force and with great velocity against the corn, but Wigfull had been in advance of his time by actually incorporating a screen and winnowing apparatus. By the middle of the nineteenth century, besides the Tasker drum, Messrs. Garrett of Leiston had a good type worked by four horses. With the coming of the steam portable engine and the traction engine threshing drums were improved, but this subject has been fully dealt with in my book *Traction Engines*. Messrs. Ransome have always made good drums and their combine-harvester of today maintains their tradition for excellent workmanship.

When I was a boy in the 1920's I well recall watching Alfred Reeve of Brook Farm, Hawstead, using what seemed to me a weird and wonderful contrivance for cutting chaff. Steam chaff-cutting had, of course, been established for many years, although at Bryers Farm I had used, in an emergency shortage, a hand cutter which boasted a knifed wheel. The one that Alf Reeve used was a very ancient implement indeed as will be seen by my sketch. Although I possess a knife from one of these, the only complete specimen known to me is in Colchester Museum.

The guillotine method used in this machine was similar in effect to the bread machine used in parish socials until the advent of the cut loaf! About 150 years ago these little "chaff-boxes" were in daily use on almost every farm. The operator had to acquire considerable knack, but as usual speed and skill came with practice and itinerant chaff-cutters (men who were often thatchers as well) would travel from farm to farm carrying the implement upon their backs. It is remarkable that these hard-labour machines continued to be used in East Anglia for so long after the knifed wheel had been established.

Fig. 26. Chaff-box

Arthur Young in his *Farmers' Calendar* (1805), page 11, says, "The number of engines (contrivances) which have of late years been invented for cutting hay or straw into chaff (most of which

execute their work sufficiently well) leaves no farmer in the Kingdom under the necessity of using the common chaff-box worked
by those only who have acquired the art of using it and who
usually make greater earnings than the common pay per diem."

Having read this extract, I marvel more than ever that 120
years later, as I have said, I saw one being used. I feel a description
of the implement (see sketch) and how it worked ought therefore
to be placed on record. It is certainly no lazy man's tool but a
mixture of simplicity and ingenuity. The "box" consisted of a
strongly made wooden trough about 4 feet long, a foot in depth
and set on four legs about 2 feet high. Hay or straw was pressed
into the trough, the fore-end of which was strengthened by an
iron facing against which worked a large knife furnished with a
handle at the top and the bottom being fixed to a movable crank.
This allowed the knife to move up and down with a saw-like
movement when in use.

Over the hay or straw in the trough was laid a thick block of
wood close up to the knife; this was to compress the hay so that it
would cut more readily. Pressure was applied by a foot pedal
attached to the block by a stout leather strap. After each cut, the
mass of hay had, of course, to move forward, the foot-pressure
meanwhile being released. At once the compression block was
raised by the action of the two-bowed springy poles affixed to the
sides of the trough. Their forward ends reached just over the block
to which they were attached by strap or wire. Once the compression block was raised, of course, the mass of hay or straw
could be moved forward. This was done with a short handled four-
pronged fork held in the operator's left hand.

In working practice, therefore, the operator stood with his right
foot on a low wooden stool and his left foot on the "pedal". With
his right hand he grasped the knife handle, the fork in his left hand.
As he drew up the knife to its highest position, ready for the downward thrust, he drove in his fork, pushing the straw or hay just far
enough for the first knife-bite. Next he would depress the pedal
and finally give a downward thrust to the knife. Again and again
would this operation have to be repeated and an expert would
make fifteen to twenty cuts per minute. From what Arthur Young
says, this job was never very easy. The splendid chaff-box in

Colchester Museum was formerly owned by the late John Owers of
Brown's Farm, Little Dunmow, who continued to use it well into
this century. In the winter of 1902, he cut in this manner with his
own hand some four tons of chaff, besides what he used for feeding
to his own stock. The knife in my collection is 24 inches in length,
and 2½ inches in breadth. The framing of the knife has its original
wooden handle, and at the other end it tapers with a forged connecting "eye" with which it was linked to the crank. These complete boxes were locally made by village carpenters and black-
smiths and sold at between 30/- and £2. Some farmworkers nick-
named the contrivance the "monkey-box" but mostly the term
"chaff-box" was used and, in fact, the self-same name is used to
denote the modern chaff-cutter.

The ability to stack corn neatly has always been a recognized
skill of the farm-worker and some fine examples of this almost
architectural skill are still seen. On the other hand the use of the
combine-harvester has resulted in many stackyards remaining
quite empty at the present day. Stack shapes are usually either
rectangular with gabled ends, boat-shaped with curved receding
ends or round. In connection with the latter, I have sketched the
old iron rick-stand at Bryers Farm upon which I have helped
Charles Long to erect many a stack.

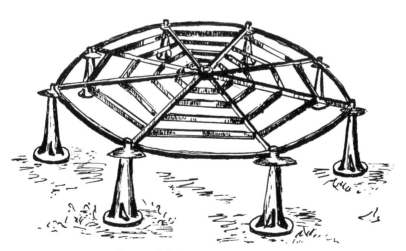

FIG. 27. Rick-stand at Bryers Farm

Its immediate use was, of course, to keep the stack off the damp ground and to prevent the harbouring of rats or mice; the "stathels" or supports were made of iron in this case, but I have very occasionally seen stone ones. Upon the framework we added wood and faggots and finally straw. We built some very high ricks on these two stands and it was hard work with the longest possible pitchforks. The granary at Bryers Farm is similarly set up on almost identical pillars to those of the rick-stand—obviously to make it more rat-proof.

The next autumnal task was, of course, thatching, and this style of work varied a good deal from district to district. Wheat-straw was always popular, and reeds were reserved for house and barn-roofs. The straw was heaped and wetted, and the thatcher's assistant (sometimes the author) had to draw out the yelms, i.e. the bundles of straw, and tie them with stick and line to be carried up the ladder to the thatcher. By the use of tarred twine and "broaches" (split hazel sticks) he would weave a suitable roof-covering to offer the greatest possible resistance to winter rains. Later on when the corn was threshed, I would often be sent to the resultant straw-stack to pull out a bundle of straw for bedding down the farm animals. I used the "scud-winder" or "throw-hook" to make a straw band with which to tie up my bunch of straw. I show both types in the photograph. The more elaborate one was given me by William Parris whose father actually made it. A sapling was bent into a curve to form a crank; at the other end is a swivel-joint like that of a flail which allowed it to turn freely, and independent of the appendage by which it is attached to the person's waist by a leather strap. His left arm is therefore freed and the spinning action soon produces a straw-rope. In the simpler variety, the hook part grasps the straw and by means of a swivel joint is made to revolve in a two-handed effort.

The land now had to be ploughed, and the first plough was nothing more than a wooden stick pointed and pushed through the soil and disturbing it to a depth of only 2 or 3 inches. The first pictorial record is found among ancient Egyptian writings and monuments, and of course ploughs are mentioned in both Old and New Testaments. The plough of the Israelites was drawn by a yoke of oxen (or an ass with an ox) and has scarcely altered

WAGON LAMPS

FOUR EARLY LANTERNS

TWO CORN DOLLIES

A HURDLE-MAKER

since the time of Moses. It hardly turns a furrow but merely scratches the soil. Nine years ago my friend, Ronald Welch, then living at Saffron Walden, was on holiday in the Holy Land. Just outside Jerusalem he discovered an Arab using one of these simple implements. He bought it and kindly had it flown to Stansted airport as a present for my collection. It has an iron share which is of sufficient size and temper to be capable of being forged into a sword. In the photograph (Plate 4) I am holding a very ancient Palestinian hoe given me by the late Mr. Percy Smith. I make no apology for including this photograph by Sydney Cross for although the book concerns British implements in general, this is an interesting prototype of *all* ploughs.

Extant drawings prove that early Greek ploughs were furnished with wheels; Cato mentions two used by the Romans, the "Romanicum" for stiff soils and the "Campanicum" for lighter land. Varro actually mentions one used for ploughing in seed, which had two mould-boards, and Pliny and Palladius make similar references. This amazes me, for I have actually ploughed in beans with a "bean barrow" and a double-furrow Ransome's iron plough. The differences in the operations could not have been very great despite a gap of thousands of years! It would seem that the ancients had the different types of plough common today, in fact, although in different forms. They had ploughs with and without mould-boards, with and without coulters, and with and without wheels. They had both broad and narrow-pointed ploughshares and it is interesting to note that a wheeled plough is shown in the famous Bayeux Tapestry and a swing-plough in the beautiful fourteenth-century Luttrell Psalter.

The ploughs of the early Britons were not so advanced, for by their law no man should guide a plough until he could make one. The Saxons' ploughs were drawn by that barbarous fashion of attachment to the tails of draught-oxen. The Normans made a wheeled plough, the driver of which also carried an axe for breaking clods. Until the seventeenth century ploughs in Ireland were drawn by oxen-tails and in 1634 an Act of the Irish Legislature was passed and I give the title of the Act herewith: "An Act against plowing by the taile and pulling the wool off living sheep". Horses as well as cattle are mentioned as being the victims of this cruelty.

Jethro Tull, already mentioned as of horse-hoe and drill fame, also did some research on British ploughs and he did a great deal to improve the implement. The Dutch were among the first to bring the plough to its present state, and their plough found its way into northern England to be later termed the "Rotherham" plough. Apart from the coulter, it was a wooden plough with draught-irons and share; the mould-board and sole were plated. Arthur Young in his Agricultural Report of Suffolk mentions an iron plough made by an ingenious blacksmith at Bradfield Combust named Brand. Young asserted, "There is no other plough in the Kingdom equal to it." This report is dated 1804 and by that time Brand had been dead some years.

In 1740, James Small began to make ploughs of an improved type in Berwickshire; but quite soon an East Anglian was to do better. In 1785, Robert Ransome (son of the schoolmaster at Wells, Norfolk) had a small foundry in Norwich and took out a patent for tempering cast-iron ploughshares. By 1803, he had removed to Ipswich where he founded what was to become a world-famous firm. Here he fulfilled his dream of a cast-iron ploughshare which would remain sharp in use by patenting his "Chilled share".

As is so often the case in a famous invention, the first one he produced partly by accident. He had been trying in various ways to produce the desired result and one day the molten iron burst out of his moulding-frames and ran about the floor of the foundry. When the metal was broken up for re-melting, Ransome found some parts to be harder than others, owing to their having come into contact with the cold iron. This gave him the main conception of the now famous chilling process. In effect, this amounted to the production of such a layer of steel on both the underside and the landside of a share, as to resist the grinding and wearing tendency of friction. A plough-bottom with mortises to receive the tenons of the wood to which it was attached, together with a movable slade, were next invented by a Suffolk farmer. These were readily taken up as were Ransome's 1808 improvements whereby he devised ploughs with movable parts all round so that repairs could be effected easily and new parts fitted.

Improvements in Scotland brought about the all-iron plough, but wooden ploughs continue to be used. I learned to plough on a Cornish & Lloyds' wooden wheelless plough drawn by a pair of horses (the subject of my memory sketch).

FIG. 28. Plough used by the author

I also won a ploughing prize in 1927 at a furrow-drawing match at Stanningfield. There is something about the "feel" of a wooden plough, a greater flexibility than with one having iron handles and beam I always think. I would like to quote here some lines by an unknown countryman which I discovered scribbled in an old catalogue:

Yes an old wooden plow and they say to be sure
The wide-awake farmer must use them no more.
They must be of iron, for wood there's no trade for.
What do the fools think God made these ash trees for?

One such plough made by Bentall's is reared annually surmounted by a huge Cross at our Diocesan Church Stand at the Essex Agricultural Show. I am glad to record also that a clergyman (long since departed) named W. L. Rham promulgated an improved mould-board which Ransome's manufactured and exhibited at the 1840 Royal Show at Cambridge. The stony land of Kent and Surrey inspired the invention of the "turn-wrest" plough, whereby the mould-board was reversible and acted either way. Other famous ploughs were the Norfolk wheel-plough with its bulky carriage presenting a clumsy appearance, although doing good work on light friable soils.

A delightful and true story concerns some modifications in plough design by a Mr. Theophilus Smith of Hill Farm, Attleborough, Norfolk. His work was noticed by the then Earl of Albemarle whose estates were near Attleborough. The Earl was then (1842) Master of the Horse and by his influence it was arranged that Theophilus should take his models to Windsor to show Queen Victoria and her Consort. He came from a sturdy family of Norfolk Yeomen Baptists, and the hour was late when he arrived at Windsor Castle. A Gentleman of the Household with the rank of Colonel received him and suggested that Theophilus spend the night at an inn. The honest farmer was indignant and replied, "that do seem mighty queer, that's both ill-convenient and costly. I was commanded to come here and the least you can do is to give me a bed. If you was to come to Attleborough my missus would find you a bed, 'specially if we'd asked you to come;

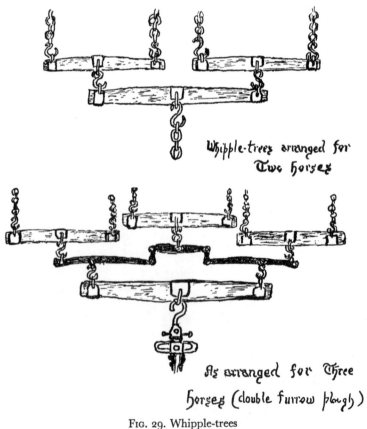

Whipple-trees arranged for Two horses

As arranged for Three horses (double furrow plough)

FIG. 29. Whipple-trees

A WOODEN CULVERT

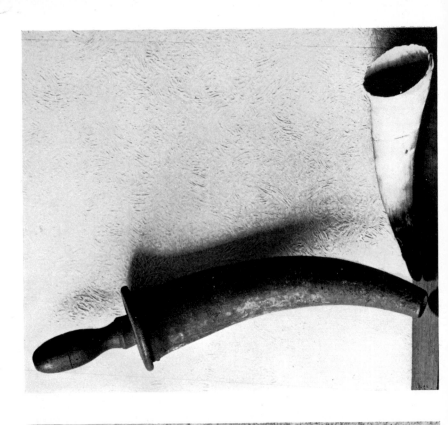

FIG. 30. Horse-ploughing

and if you was as hungry as I be, I warrant she'd find you something to eat."

The Colonel was decent enough to find him lodging at the Castle and next day the farmer was ushered into the Royal Presence. First of all it was Prince Albert who allowed the plough to be called after him, then came the Queen herself. Theophilus afterwards wrote down his astonishment. He had expected a person "with a gold sceptre in her hand and her gown all a-trailing behind, as we see in the picters; but there she was a comely simple woman with a kind look". They talked of ploughs, wages, cottages and the poor. "By and by", said Theophilus (afterwards telling the story to a friend), "I begun to get uneasy—'Theophilus,' says I, 'you're brought before Princes and you must testify.'" The Queen gave him an opening: "How did you come to think of this clever invention?" she asked.

"Well your Majesty," he replied, "I had it in my head for a sight of days before it would come straight. I saw what was wanted well enough but I couldn't make out how to get it. At last I made it a matter of prayer, and one morning it came like a flash." "Why, do you pray about your plough?" asked the Queen. "Well there Mum, why shouldn't I?" he replied. "I mind one of my boys when he was a little mite, I bowt him a whip and rarely pleased he was. Well, he comes to me one day crying as if his heart would break. He'd broken the whip, and he browt it to me. Well now your Majesty, Mum, that whip was nothing to me, but it was something to see tears running down my boy's cheeks, so I took him on my knee and I comforted him. 'Now don't you cry', said I. 'I'll mend your whip I will so that it will crack as ever.' Well now, don't you think our Father in Heaven cares as much for me as I care for my boy? My plough wasn't of much consequence to Him, but I know my trouble was." Thus it comes about that Messrs. Ransome's records contain an invoice dated 22 June, 1842, for £7 14s 0d made out to T. Smith of Attleborough for 1 Patent "Albert" Plough with gallows, wheels, coulter and one dozen shares on behalf of the Queen and directed to H.R.H. Prince Albert of Windsor Castle.

Some ploughs made by Howard's of Bedford had both handles and beam in one piece with a draught-chain attached to the body. Double-furrow horse ploughs were in use for a long time. The coming of the steam plough has been dealt with in my book *Traction Engines*.

Spud

Spanner

FIG. 31. The Ploughman's tools

I would like at this stage to include Geoffrey Chaucer's description of a fourteenth century ploughman. "There was a ploughman that had driven full many a load of dung. He was a true and good worker, living in peace and perfect charity. He loved God best at all times, whether he was happy or distressed, and afterwards he loved his neighbours as much as he did himself. He would thresh, dig and make ditches in the name of Christ for any poor fellow without pay, if it lay in his power. He paid his rents promptly and in full, both with his work and his cattle."

I am proud to say that I have known such men in my own day and generation.

I would like to mention a breast plough which Percy Wedd of Audley End gave me for my collection and a photograph of which is reproduced. This hand-pushed plough was used in paring turf and for actually ploughing small acreages. The face is angular and sharp, the blade is worn but measured 15 inches by 9 inches, the right-hand straight side being turned up 3 inches with a cutting edge in front. The flat helve is 5 feet long with a cross-handle; this is raised to the height of a man's haunches for pushing. I recall seeing modified versions of this breast plough being used in the allotments and gardens of Culford in the 1930's.

Fig. 32. Elephant Plough

WINTER

WINTER

CHAPTER SEVEN

THE contemplation of winter on the farm at once brings the barn into focus. Our British barns have always been big enough to contain the produce of farms; when the rickyard became more popular, however, barns became smaller. Today, with the growing use of the combine-harvester, the grain-silo is replacing the barn. The parson's tithe-barn years ago was always important and some fine architectural specimens are left to us. Many ordinary boarded barns however, have an architectural magnificence and with their naves, aisles and transepts are not entirely unlike churches. At Bryers Farm, the stable barn was early seventeenth century, but many of the crossbeams and queenposts had been used before, judging by their mortise holes and notches. It is not easy therefore to estimate the age of the actual timbers. The older barn would contain the specially prepared threshing-floor referred to in a recent chapter. Parson Rham has left us a recipe for a threshing-floor. "The soil is taken out to the depth of six to eight inches or more, and if the subsoil is moist a layer of dry sand or gravel is laid at the bottom three or four inches thick, and trod smooth and level. A mixture is made of clay or loam and sand with water to the consistency of common building mortar; to which is added some chalk or pounded shells or gypsum. Chaff, cow-dung and bullocks' blood are added and the whole is well worked up together. A coat of this is laid on the prepared bottom with a trowel, about an inch thick and allowed to dry. Another coat is added and the cracks filled up carefully. This is repeated until the desired thickness is produced. When it begins to harden, the whole is well-rammed with a heavy wooden rammer and every crack filled up, so as to give it the appearance of a uniform solid body. This is left to harden slowly and in a short time becomes sufficiently hard to be used."

We have already dealt with threshing, which operation was

always followed by the well-known winter job of winnowing and cleaning the grain. The very earliest method was merely to toss the grain into the wind, but this inefficient method was superseded by machines quite early in history. The Chinese brought out the first apparatus for this purpose in their ricefields. The scheme reached Scotland where Andrew Fletcher and James Meikle worked on the problem. The usual principle was a fan rotated by a geared handle. My brother at my suggestion gave our very fine old specimen to Reading Museum some years ago and through their kindness a photograph appears on plate 6 which shows the beautifully handmade iron flywheel. My father and I often used this machine for clover-seed. Subsequently riddles were added to similar machines called "dressing-machines" and sometimes the larger types were fitted for horse-power or water-power.

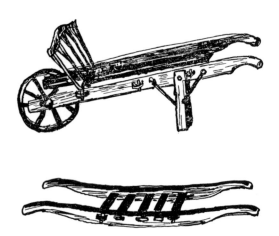

Fig. 33. (*top*) Sheaf barrow
(*bottom*) Hand barrow

A host of small hand implements are associated with the barn, and I have sketched a group of the most interesting specimens. Other machines to be found in a barn were various types of mincing machines for mangolds and turnips. The earlier type was a simple frame with a guillotine cutting knife similar to the early chaff-box. Oil-cake breakers were also known well over 100 years ago. There was for instance the corn barrow which carried sheaves

wicker-work were used in East Anglia, subsequently to be re-
placed by tin or wooden skeps. A wooden hoe about 7 inches long
and 4 inches deep in the blade with a 9-inch handle was a useful
thing for drawing in grain to fill the skep quickly. Wooden shovels
or scoops were indispensable, and were only 3 feet 3 inches high,
with a handle like that of a common spade. The blade, helve and
handle were in one piece, the belly of the scoop being a little
hollowed out and its back thinned away to sides and face. Corn
was always sold by the bushel and levelled-off with a strike in those
days. The strike (of which I have a good specimen) is drawn
straight across the imperial bushel for true measure. Sack-barrows
do not seem to have changed in the years but several of us have
sustained back-injury through corn-carrying years ago. This has
been mechanized today and there is a move to do away with the
big four-bushel sack. Near the barn door at Bryers Farm notches
were cut where the sunlight streamed in and this made a crude
sundial which enabled the workers of a former generation to know
the time of day.

Tally-sticks were the method of recording hours of work, and
really were as accurate as the clocking-in methods of any modern
factory. Two sticks of similar length were laid side by side and a
notch made by the bailiff or "lord" across both sticks, according
to the number of days or hours of work done. One stick he re-
tained and the other was kept by the worker himself. In cases of
dispute, the two tally-sticks were again laid side by side. If the
notches did not agree, i.e. if one had more or less notches than the
other, somebody was not being honest. Many rural shopkeepers,
tradesmen and craftsmen made similar use of tally-sticks in all
their business transactions. Before the days of cricket scorebooks,
all scoring at cricket matches was accomplished by cutting notches
on sticks.

Weights and measures in earlier times did in fact cause con-
fusion. Our rod, pole or perch is now fixed at $5\frac{1}{2}$ yards but on the
old manors the tenants' perch was reckoned as 18 feet and the
landlord's 18 feet 6 inches. Even the bushel varied according to
which grain was being measured, and a bushel of oats was not
equal to a bushel of barley. Horse measurements even today are
reckoned in hands, and for years I have practised making steps of

FLAIL BASKETS, COSTRELS IN LEATHER, WOOD AND STONE, AND HORN CUPS

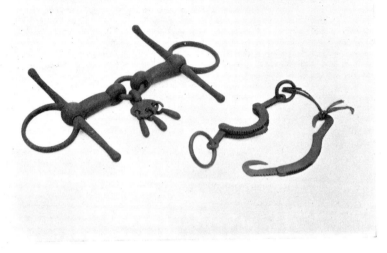

THREE BITS

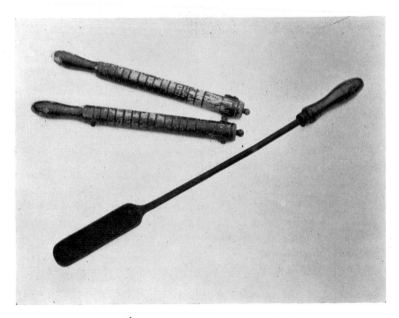

HORSE'S MOUTH-CRAMP AND TOOTH FILE

a yard long in spite of the fact that I wear a size 14 boot! Like many more countrymen too, I can measure the length of fences pretty well by remembering that a person's height is nearly always the same as the span from finger tip to finger tip if the arms are outstretched sideways. If I want to measure in inches I always use the breadth of my thumb.

I have been shown a wages sheet for Essex dated 1661 which makes an interesting record: Mowing one acre of grass 1/10. Raking and cocking one acre of grass 2/-. Reaping, shearing and binding one acre of wheat 4/-. Ditto rye 4/-. Mowing one acre of peas or vetches 1/9½ and threshing 2 coombs of wheat 1/-. Most of the work was then piece-work.

Sir John Cullum in his *History of Hawstead* dated 1784 has some interesting disclosures to make about farming in his day; he says, "The farmhouses are well furnished with every accommodation; into many of late years a barometer has been introduced. The teapot and mug of ale passes jointly the breakfast table; and meat and pudding smoke on the board every noon. Formerly one might see at Church what the cut of a coat was 50 yrs. before: No such curiosity is now exhibited, every article of dress is spruce and modern. A head servant-man lives in the house and gets from seven to eight guineas a year. A maid gets three, and a boy two guineas. A day labourer gets 1/2 a day in summer and 1/- a day in winter plus a beer allowance. A man gets paid 1/- for threshing a coomb of wheat, and 6d or 7d for oats or barley. For mowing grass he will get 1/4d an acre and a 'weeder of corn' 6d per day."

Corn was taken to Mark Lane and hay to the Haymarket in London from East Anglia by horse-drawn wagons until this century. So accustomed did the horses become to the journey that the carters would often sleep on the loads. So little was the traffic that safe arrival was usual. At a later date it became compulsory to carry wagon lamps and I have a lovely old set of these, which came from Sam Middlebrook of Hempstead. These are rear lamp and two side lamps whose lighting power was candles (see photograph). Geese and turkeys from Norfolk, Suffolk, and Essex were driven up to Leadenhall and Smithfield Markets by road. My father has told me of the method by which their feet were shod for protection on such a long journey. They were driven over warm tar and sand which formed a protective coating to their feet. Less than 100 years ago this was quite a common practice.

With the continued increasing use of pig-wire fencing and electric fencing it would seem that the days of the hurdle-maker are numbered. An early boyhood memory takes me into Bryers Wood in winter watching Charles Elsden of Whepstead and Jim Smith of Lawshall in their improvised huts making hurdles from ashpoles and broatches (thatching-sticks) from hazel stems. My father would then chalk out a notice on the tarred side of the roadside stable "Hurdles and broatches for sale" and at an early age I would be delivering them by horse and tumbril. Jim Smith protected his knees with leather knee-caps and his work lost him the sight of an eye in a hurdle-making accident. Later on as a farm student at Chadacre, I got to know "Banty" Hale who worked there. William Rickard, a recluse from Littlebury Green, made hurdles until his death a few years ago and Philip Taylor (whom I photographed) still does the work at Audley End.

About 100 years ago these itinerant hurdle-makers made new hurdles at 4d a piece and provided their own tools. I have some of these tools (see photograph) which I rescued from an old iron depot at Chelmsford, by the kindness of Mrs. Fairhead of Riverside, who also gave me the scissors shears referred to in Chapter 2. They were also used for rending oak laths by builders. Willow was a popular hurdle wood, but the men I knew used ash poles. Occasionally a shepherd made and mended his own hurdles. The tools consisted of a handsaw, a light hatchet, a draw-shave, a riving-iron or flamard, a centre bit, gimlet, hammer and auger. The latter was sometimes called a "wimble-wamble". In addition there would be a home-made rending frame, which in reality was a trestle on which two strong poles were laid leaning and connected by a piece known as a bridge. A stem of a tree or a post let into the ground was needed for shaving the poles. In such a post, two holes were bored with the auger to admit two stout iron stubs which projected for six inches. These formed a vice to hold the poles whilst they were being shaved.

Placed three feet away from the post (or stump) was a standard having a spike at the top which steadied the pole when the draw-shave was being used. Into the same post was driven a square staple to hold the bases of the heads when they were being mortised. First of all, the butt end of the pole was shaved off; the pair of heads were 4½-feet lengths. Nine-feet lengths made a pair of slots and 5-feet lengths made a pair of stay-slots. Uprights consisted of 3½-feet lengths and the different connecting pieces were made on the rending-frame. This was a fascinating process. The piece was put over the bridge with the butt-end upwards, the rending-iron (flamard) was then placed across the pith and driven down by a wooden mallet. After the first two feet, the pole was borne on to the bridge while the iron was guided to make an exact cleavage. All the different pieces were thus "cloven". A bit of chopping with the axe was necessary, to remove bark or other irregularities.

The next procedure was to form the actual hurdle. Four low stumps were driven into the ground to mark the length, and four others to mark the distance between the upper and lower slots. A pair of heads, one at each end, were laid down in their right position, and the six slots were laid at the correct distance duly "soubed" to the size of each slot to regulate the mortises. The joining was then begun with holes drilled at each end of the mortises as well as one for the diagonal brace slot. The mortises left an intermediate piece between the apertures and this was taken out by the "tomahawk", a tool made for this purpose. Top and bottom slots were next nailed to the heads. For every nail used, a gimlet hole was first made to avoid splitting—the clenching of these nails finished the operation. The nails used were of "best iron" fine-drawn (not square) but flattened to facilitate clenching. These nails years ago were bought at 6d per lb.

All this sounds very tedious but under the expert hands of these craftsmen a dozen beautiful hurdles would be made in a nine-hour day, providing the material was clean and pretty straight. One hundred poles made about three dozen hurdles which with labour and nails cost about 10/6 a dozen. The average hurdle was 8 feet long and very strong indeed. Eight of these to a chain of 22 yards was allowed. Shepherds and their "pages" who helped them were adept in using the finished article, and I have watched many a wagon-load taken on to a field where in a short time, Joe and Percy Mason would erect a fold on a patch of turnips or kale. The hurdles were laid down flat, end to end at first, and then a

big dibber called a "fold-drift" was used to make a hole for the
pointed hurdle-stake. An iron shank was passed over the heads of
the hurdles for additional strength—after all, the enclosure was to
hold a flock of sheep within bounds!

On large estates, sometimes bigger hurdles termed "park
hurdles" were used for subdividing meadows or pastures. They
were sufficiently strong to form a fence against cattle. The ordinary hurdles were put to a variety of uses quite apart from the sheep-
fold. A gap in the hedge would be stopped by a hurdle, a wind-
break for poultry or young calves or pigs was made by "thatching"
one or two hurdles with straw. When loading pigs for market, a
man holding a hurdle could and did exercise far more control
with than without it. More than once a hurdle has carried off a
person when having had an accident on farm or hunting field.
Lastly, a certain small boy—who shall be nameless—fired with
the Robinson Crusoe idea, would make a fine tent by stretching
canvas over a couple of hurdles leaning toward each other.

CHAPTER EIGHT

Winter time always brings reminders of the portions of any farm which need draining. This is the withdrawal of superfluous and injurious moisture from land by means of artificial methods. Simple methods of so doing were known generations ago. Cato, Palladius, Columella and Pliny all write about draining and seem to have been familiar with its advantages, as well as being able to construct both open and covered drains. Pliny in fact mentions some methods which 100 years ago were hailed as new inventions. My late father used an expression which I think cannot be bettered in describing land which needs draining—"water-slain"; and excessively wet land is most certainly dead land. On undrained clay soils, seeds such as wheat simply rot in a wet season. The innings of Romney Marsh and the fosse-dyke in Lincoln along with many ancient lodes and drains, were the work of Roman Legions. The marsh folk of Kent and Sussex had their own method of controlling water levels in ditches and dykes. They made a wooden culvert from the trunk of an oak tree hollowed in its centre; one end being left open and the other end blocked. At the blocked end they fitted a wooden tapered plug shown in the photograph (by Mr. Douglas Weaver of Ashford). The trunk was laid between two drainage channels lowered into the ground so that the farmer could control the water level from one ditch to another by simply removing or replacing the plug. This magnificent specimen was found by Mr. Alec Douglas of Tenterden whilst re-draining land between Newenden and Rolvenden on the Kent-Sussex borders in March, 1960, and is said to be one of the best specimens ever discovered. The subsequent work of the Dutch in the English fen country is also well known. About the middle of the seventeenth century a book was written by Captain Walter Blith which was in advance of his time and its general excellence was not fully appreciated until 100 years later. His methods were practised in Suffolk, Norfolk, Essex and Hertfordshire, with beneficial results. The old method of subsoil draining of arable land consisted merely of a series of simple conduits at shallow depths

73

dug across fields. Another old form was bush draining and I re-call, soon after having qualified at Chadacre Agricultural Institute helping Charles Long to lay one of these drains at Bryers Farm. We dug out a deep trench, and cleaned it with crumb-scoop. The digging was by a narrow draining-spade, but in my collection I had two very much older types and I sent one of these to the Science Museum. Also shown in the photograph is a wooden spade used in the fens 120 years ago and given to me by Sam Middlebrook of Hempstead. A subsoil fork from Geoffrey Mason has since been added and is shown on another plate. When we were deep enough we pushed in brushwood (mostly ash) and pressed this down tightly with a forked stick. It seems that the astringent nature of such wood preserves its substance from decay under ground for a considerable time. Anyhow we filled in this trench and I see it running copiously whenever I revisit the farm. Such drains of course could only be satisfactory in stiff clay soils. In other districts tiles and stones were used instead, and of course the porous drain-pile was a natural follow-on. Later still, mole-ploughing came into being, using a draught implement called a mole-plough with which a circular duct of $2\frac{1}{2}$ to 3 inches in diame-ter was made at a given and uniform depth from the surface. This was formed of course not by any extraction of earth but entirely by compression, and the whole of the drain except a slit along the top is so firm and impervious as to resist the infiltration of water and to act solely as a conduit pipe. The boring instrument was a cone-shaped share or bolt of $2\frac{1}{2}$ to 3 inches in diameter at the back and sharp at the point. This was attached to the beam of the plough and held fast by a strong wrought-iron bar. These were horse-drawn but thanks to the work of pioneers like John Fowler the job was mechanized and the steam engine used. This has of course been covered in my book *Traction Engines*.

Hedging and ditching were always winter jobs, and about 100 years ago there were two classes of hedger on most farms. The "superior" hedger ranked with the master-ploughman and was a man of great value and intelligence. He not only had charge of the pruning and the plashing of hedges but also the planting of new ones. The "inferior" hedger ranked with the under-plough-man and only a large farm or estate would keep a "superior" hedger at all. The hedger's tools were simple but very effective. A normal smallish spade, for ditch work and any shrub planting, was an absolute necessity. Other tools were a hoe, one or two hook-headed sticks and a hedge-bill (see sketch). The latter consisted of flat, broad, curved pointed blades of iron hardened into a cutting edge and terminating in a split socket into which was fitted the helve or handle made of ash. The other requirement was a smaller slasher for lighter work, and, of course, a light axe and stout leather gloves. In recent years, I discovered that workers on the Audley End Estate near Saffron Walden, always use an old sword for light trimming.

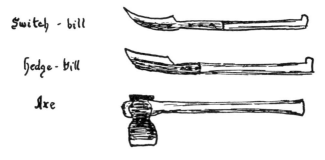

Switch - bill

Hedge - bill

Axe

FIG. 35. Hedger's tools

Hedge-laying and plashing is a very ancient method whereby old and neglected hedges are ruthlessly cut and the gaps filled in by plashing. In this operation, the stout stem is cut fairly deep and bent down, fastened by a hooked branch of a neighbouring thorn or by pegging down to the ground by a strong hook-headed hook peg. A wedge-shaped stick is pushed into the cut to prevent the bent bough from starting up again. Usually, some clay was daubed on the incision to protect the severed tissues of the wood from too severe action of rain or sunshine. A good fence can often thus be made. In these days, hedges of this type have become less and less fashionable, and concrete or wooden posts and wire, together with electric fencing, are more popular. Nevertheless, I am recording a method taught to us as students at Chadacre Agricultural Institute in 1922-4 by Philip Hammond, an expert hedger from Shimpling. The art of hedging did not merely consist in laying these plashes so that they would not snap away from the stool, but in selecting the right boughs to lay down. Phil Hammond would contemplate and judge the lengths and angles of the stems and allow for which type of wood or thorn he was using. His work was all done in sections. No doubt different districts had different methods but it would be slightly beyond the scope of this book to describe them all.

The practice of manuring fields for greater crop yield is by no means new. Homer mentions in a delightful way an old king who was found manuring his fields with his own hands, and also describes a dog lying upon a heap of dung with which the labourers were about to manure the farm, and many of us have seen a nice well-bred clean dog do exactly the same thing. Augeas is celebrated as the discoverer of the use of manure in Greece. Xenophon represents earth which has long been under water as a fertilizer of the soil and recommends leguminous crops to be grown for the purpose of ploughing-in as manure, and remarks of such that "they enrich the soil as much as dung". Virgil indicates some acquaintance with the ameliorating effect of a change of crops, suggestive of the Norfolk tour-course rotation of crops brought in centuries later. Virgil recommends *nitrum*, not saltpetre as most translators say but carbonate of soda or potash, in mixture with the dregs of oil as a steep to make the seed-grain swell. He also suggests the value of scattering ashes over the land, and even mentions the value of pumice stone and shells.

Cato, Theophrastus and Columella write fully on this subject, and they display a knowledge of organic, inorganic and mixed manures which would have done credit to distinguished British farmers of the sixteenth century and indeed much later; in fact, the use of farmyard manure appears not to have varied from Roman times until a hundred years ago. A seventeenth-century writer spoke of the value of snow as equal to the richest manures, impregnated as this is with "celestial nitre". Guano was esteemed in Peru generations ago, but was not commercially introduced to Britain until about 1840. Crushed bones were likewise unknown in Britain till that time. A remarkable prophecy was uttered by Liebig when he wrote, "A time will come when plants growing upon a field will be supplied with their appropriate manures

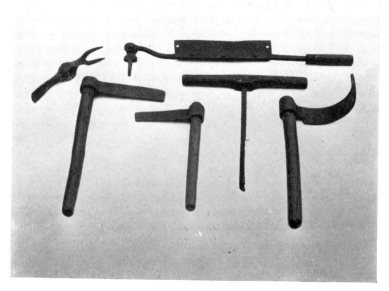

TURNIP CHAT-HOE, TURNIP CHOPPER, AUGER AND RENDING IRONS

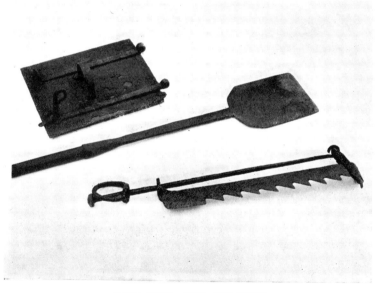

OVEN DOOR, PEEL AND HAKE

BUTTER MEASURING IMPLEMENTS

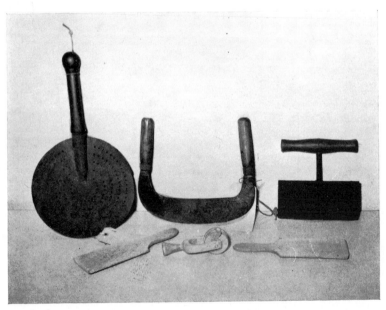

BREWING-SKIMMER, CHEESE CUTTER, SUET CHOPPER,
BUTTER-HANDS AND MOULD

prepared in chemical manufactories, when a plant will receive only just such substance as at present a few grains of quinine are given to a patient afflicted with fever; instead of the ounce of wood which he was formerly compelled to swallow in addition."

One of my hardest farm jobs was to fork out and throw on to the tumbril hard well-trodden farmyard manure resultant from a bullock yard in which animals had spent a whole winter and which manure was well and truly trodden. I have sketched the tools we used for that purpose. Out on the field it was either heaped or applied direct being raked off the tumbril into heaps by use of a muck-crome.

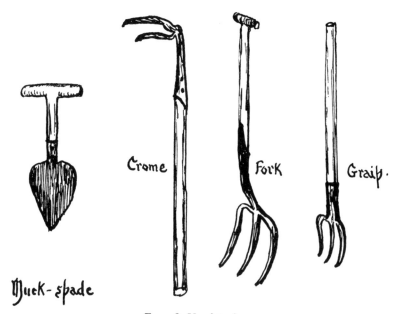

Muck-spade   Crome   Fork   Graip.

FIG. 36. Yard tools

Early artificial manures were simply scattered by hand from the seed-lip, already mentioned in Chapter One in relation to drilling. When the urine of animals could be collected, a liquid-manure cart was used to take the liquid direct to the fields. Often this consisted of a cask of 120 to 140 gallons capacity, suspended between the shafts of a cart, which meant that the axle had to be bent to nearly a semi-circle. The cart in fact was a mere skeleton consisting of 14-feet long shafts connected fore and aft by a crossbar placed at such distance as to allow the length of the cask, and the width between the shafts was suited to the diameter of the cask. A pair of rather broad cart-wheels were fitted to the axle; and a funnel-shaped hopper was placed over the bung-hole. The distributor section was made of copper, cast-iron, or even wood, and extended from wheel to wheel, being perforated for regular distribution of the liquid. The distributor section was attached to the cask by means of a stem of the same materials and bore as the perforated main tube. I have tried to sketch this implement which was often adapted as a farm water-cart. Steel-tank carts were subsequently used for liquid manure but the whole practice seems now to have passed away.

Fig. 37. Cart for water or liquid manure

Machines for distributing dry fertilizer and bone meal have been mentioned in Chapter One (drilling). A very good type was invented by Farmer Chambers of Fakenham and made by Messrs. Garrett of Leiston. It is so like a corn drill in build, shape, and action, that it would seem wrong to describe it in detail. The mechanization of manuring with farmyard manure is an accepted fact today but belongs completely to the era of the tractor, and the manurial value of sheepfolding was recognized a long time ago.

I have spoken of the importance of the work of the country

carpenter and wheelwright. These men made the carts and wagons, shod their wheels and repaired the implements and farm buildings. The village blacksmith has almost disappeared and is now termed an "agricultural engineer". The old time forge (frequently sited literally under a spreading chestnut tree) was a meeting place for all ages. I have spent hours watching Harry Mortlock at Hawstead shoeing horses. When he ceased to function we went to Huffey's at Sicklesmere. He was a member of a famous family of smiths who were also adept at wrought-iron work and made and repaired all types of farm implements. Happily this craft is still maintained and demonstrated at many British agricultural shows. A shoeing-smith today who qualifies as a master farrier has to pass a quite formidable examination in horse anatomy and disease. In early Victorian days, a smith was often a self-taught amateur veterinary surgeon! I have a nice specimen of a bar-shoe sent me by Mr. M. T. Brooks of Lakenheath, and specially adapted for a tender foot. It is designed to protect a sore point from pressure, by causing the whole weight of the limb to be borne by the other portions of the shoe. The chief feature is a continuation of the common shoe round the heels. In cases of sand-crack or pumiced feet this was invaluable. I also have a set of fourteenth-century palfrey shoes sent me by Miss Ivy Pask of Hartest where they were found at great depth when digging was in progress. I also have three interesting "bits", two of which are rather brutal. The other represents the heavy type used for breaking in a horse and is fitted with a tiny hanging "comforter" which was said to tickle his tongue to keep him sweet-tempered. The other two are sharp with serrated edges which I would think could ulcerate the base of the mouth and even tear it apart. Extreme bits of this type (see photograph) are sheer wanton cruelty. My father bought them in a job lot at an auction when I was about seven years old but I can honestly say these were never used at Bryers Farm. I also show a file for horse dentistry and a cramp for holding open an animal's mouth when dosing it. This was given me by Bill Wright of Wethersfield. Today, with fewer horses in use on farms, it is pleasing to record some farmers who believe in them. Mr. Raymond Keer of Woodbridge is Secretary of the Suffolk Horse Society and tells me that every Suffolk Punch now

living traces its descent in direct male line in an unbroken chain
to a horse foaled in 1760 and bred by a Suffolk farmer named
Crisp, and the actual history of the breed goes back to 1506.
Suffolk breeders have never aimed at producing a horse with a
massive spreading hoof but have devoted their attention to breed-
ing an animal with a sound medium hoof and a clean leg. Docility
and tractability are features of these lovely chestnut animals. The
late Mr. S. G. Buck of Stowmarket had broken in about 300 of
this breed. Fine prizewinning examples may be seen at Mr.
George Colson's Long Melford Farm and also at Mr. Harry
Hall's at March. Both these farmers use horses and horse-drawn
implements. Others doing likewise with Suffolks are Messrs.
George Sadler, Whittlesford; P. Adams & Sons, Felixstowe; R.
Cooke & Son, Yarmouth; Col. Sir Robert Gooch; A. W. Hewitt,
Norwich; F. Newton Pratt, Trimley; A. H. Worth, Soham; J. W.
Bullard, Beccles; G. C. and F. C. Knight, Melton Constable;
J. W. Kemp, Diss; W. Mackie & Sons, Sibton; W. C. Saunders,
Billingford; R. T. Dawson, Norwich; G. M. Chrystal, Thurlow;
Philip Woodward, Old Newton; Guy Blewitt, Boxted; D. Cross,
Ixworth; E. Kayler, Bradfield; Alec Steel, Prittlewell; and Messrs.
Truman, Hanbury & Buxton, the brewers. The massive Shire
horse is likewise by no means a thing of the past, and is popular in
the fens. Mr. Cole Ambrose near Ely had about 600 Shire horses
on his estate until recently. Mr. Frank Starling works about
twenty-six horses at Littleport and Mr. Leonard Childs keeps a
number at Chatteris. Others using Shires are Mr. A. R. Bradshaw,
Manea; Denys Benson, Chelmsford; Charles Clark, Maidenwall,
Lincolnshire; J. A. Forrest, Gainsborough; I. & W. Whewell,
Radcliffe, Lancashire; Francis Bowley, Loughborough; Andrew
Brothers, Barton on Humber; G. Brownlow, Wisbech; H. E.
Robinson, Higham Ferrers; A. L. Lester, Leckhampstead; J. R.
Faulkner, Aylesbury; G. T. Ward, Wisbech; J. H. Martin & Sons;
Herbert Ward, Emneth; A. R. & D. Clixby, Pilham; E. J. Richard-
son, Moreton-in-Marsh; A. H. Clark, Spalding; J. H. Frank,
Northampton; J. W. Vaughan, Nantwich; H. Sutton, Lymm,
Cheshire; and Messrs. Whitbread, Mann, Crossman & Paulin,
Young & Company, Courage and Barclay, all famous London
brewers.

Clydesdales are well represented in the north of England. Among their enthusiasts are Messrs. David Batey, Penrith; S. Brooke, Doncaster; A. J. Woodward, Maulden; Irving Holliday, Penrith; James Kilpatrick, Wigton; J. Parsons, Ambleside; A. W. Seaman, Halesworth; Mrs. Colson, Melford; T. C. Alderson, Northallerton; William Weir, St. Albans; the National Coal Board at Ashington; several Scottish farmers, and John Drennan of County Derry.

Among the Percheron breeders are Messrs. Chivers, Histon, Ison, Cambridge; J. H. Martin, Downham; Alfred Lewis, Thetford; Allan Alston, East Tuddenham; James Alston, Wymondham; Geoffrey Peacock, Morley; C. O. Rushby, Helpringham; G. E. Sneath & Son, Spalding; Turner's of Soham, Ltd.; Garner's of Willingham; F. G. Starling, Littleport; Vaux Breweries, Sunderland; H. V. Alden, Burgh St. Peter; A. S. Rickwood, Chatteris; E. Bailey, Willingham. In the Cambridge area, the Frederick Hiam Farms, Ltd. keep all breeds and use them extensively, as do Mr. John Foulds of Epping, Mr. Wilson of Littleport, Messrs. Burton & Sons of Haddenham, and G. E. Bedford & Sons, Littleport.

As mentioned in Chapter Two, in the London area we still hold the Annual Whit Monday Carthorse Parade, and although entries have greatly lessened yet there are large numbers still kept. The big brewers have in fact been among the horse breeders' best customers. The London Cart-horse Parade Society was formed in 1886 by the first Sir Walter Gilbey and Baroness Burdett-Coutts. The former's grandson, Mr. Walter Gilbey, serves on our Committee today. The Society's objects have never changed—i.e. to "improve the general conditions and treatments of London's Cart-horses; to encourage drivers to take a humane interest in the animals under their care; and to encourage the use of powerful cart-horses for heavy work on the London streets". The photo (by Van Hallam) was taken at the 1959 Parade; the Society owes much to Mr. C. R. Hannis, Mr. John Young, Mr. T. A. Dobie, and its Secretary, Mr. A. G. Holland. I have already mentioned some of the brewers. Other firms doing likewise are coal-merchants and carters, i.e. Messrs. A. C. Lloyd, Ltd., Rickett Cockerell & Co., Mark Vice, Ltd., G. Hinchcliffe & Co., Turner Byrne & John Inns, Ltd., L. Newell, W. H. Ford, W. C. Hill, T. W. Farmer,

V. Burden, and the Hurlingham Club. This book is of course con-
cerned with old farm implements but I trust I will be excused for
alluding in some detail to the magnificent horses who drew them.

Agricultural shows, i.e. public exhibitions of farm-stock and
implements, have done a great deal to encourage progress among
implement manufacturers. This was especially true in the early
days of the Highland Agricultural Society of Scotland founded in
1784. Ten years later the English Board of Agriculture and In-
ternal Improvement was established and had as its President Sir
John Sinclair, and Arthur Young was its first Secretary. Subsequently the Royal Agricultural Society of England was formed
and held its first show in 1839 at Oxford, as mentioned in Chapter
Three. More than ever the present-day Smithfield Show caters for
machinery, but at first it was practically a fat stock show pure and
simple.

CHAPTER NINE

It would not be out of place to mention the smocks worn by farm-workers; the custom of wearing a smock is of great antiquity. I suppose we borrowed many of our early fashions from France and in an extremely rare book *Livre d'Heures* printed in Paris in 1504 there is a picture of shepherds listening to the Angelic Choir, and they are shown in the costume of the fifteenth century wearing belted smocks. The discontinuance of belts in England was partly because wearers liked them loose and also in consequence of one of the sumptuary laws which ordered labourers not to wear girdles unless made of linen. The embroidery was a copy of that found on women's smocks since Saxon days. The Carpenter's Wife in Chaucer's *Canterbury Tales* wore a white smock and smocks were among New Year gifts made to Elizabeth I. Two generations ago a farm-worker's smock was often called a "Gaberdine". Shakespeare mentions it as such worn by Caliban in *The Tempest*. Often the material used was drabette and in the time of Sir Roger de Coverley farm-workers wore white smocks on a Sunday. I have two nice specimens, one a Suffolk cowman's smock (see photograph) and another with more embroidery, a Lancashire Carter's smock. The smocks lasted for years in all weathers and were worn with either corduroy or fustian breeches. The full width of material being gathered into folds by smocking made extra protection on back and chest just where it was needed. Shepherd's smocks were longer, extending well below the knee and there is a lovely specimen in Saffron Walden Museum. In summer time a straw hat with a wide brim was typical farm head-gear and known as a "Zulu".

The old time farm-worker took his food into the field in a woven rush basket called a "flail basket". I have a nice specimen (see photograph) and its duplicate I sent on to Reading Museum. It was big enough to hold a rabbit if its owner was lucky enough to kill one in his long day in the fields. His drink was taken in a costrel or keg (see photograph). My earliest specimen is a leather bottle, then came the delightful little wooden tubs shown in the

83

picture, a later development being the stone bottle, also shown. The wooden costrels were hung on the hames or tees of a horse-collar at haytime and harvest tide. One poured from them into horn drinking-cups (also in photograph). The meal breaks had characteristic names in those days—the eleven-o'clock snack was "elevenses" and at four o'clock "fourses" or as Charles Long used to term it "Bever". The latter is an interesting term and in Marlowe's *Dr. Faustus* old Gluttony complains of his lack of food saying, "My parents are all dead and the devil of a penny have they left me but a bare pension, and that is thirty meals a day and ten 'Bevers'—a small trifle to suffice nature." The word must surely come from the same root as beverage—Old French in fact termed drink as "beivre". The use of horn was varied. A hollowed horn was used to give animals their medicine and I have two nice types (photographed herewith). Among the range of lanterns I have a fine one with horn sides instead of glass—a genuine lant-horn given me by Mrs. Ilott of Littlebury Green (seen on the right of the photograph). The tanner who purchased the hides of cattle separated the horns and sold them to the makers of these lant-horns and combs. The horn consists of two parts, an outward horny case and an inward conical-shaped substance. The first process consisted of separating the two parts by means of a blow against a block of wood. The horny outside was then cut into three portions by a frame-saw. The lowest portion was made into combs, after several flattening processes. The middle of the horn, after being flattened by heat, and its transparency improved by oil, was split into thin layers which formed the glass substitute for lant-horns. The tip of the horn was used for knife-handles and the tops of whips. The interior cone was boiled down, and the fat used for yellow soap. The liquid itself became glue, the chippings were used for manure and the shavings from the lant-horn maker were cut into figures and painted as toys as they curled up and became animated when placed in the palm of a hot hand.

My grandmother and my mother were both adept at bread-baking, and as a boy my Saturday morning job was to cut faggots in half to "fire" the brick oven. I reproduce a photograph of the oven door and the "peel", but the poker or "rashling pole" has long since disappeared. The dough was set some hours before (sometimes overnight) in either earthenware pans or the kneading trough. A stone of flour was taken, a little salt added, then a tablespoonful of yeast mixed in a little warm milk. After mixing with skim-milk from the dairy it was kneaded by hand. It was covered with a warm cloth and then left to "rise". This process was called "laying the sponge" and the final job was to cut the mixture into portions and place in the oven. Years before this time when few of the poor had ovens, the squire's or farmer's ovens were available for use at certain times denoted by the ringing of the oven-bell in the church tower.

My father brewed three times a year, and I used to fetch water from the well to the ancient brewhouse behind Bryers Farmhouse. I still have his old skimmer (see photograph). I give his recipe for home-brewed beer which was never denied to visitors and workers, whilst it was a most popular drink at haysel, harvest and threshing times:

2 bushels of malt ⎫
1 stone of sugar ⎬ 50 gallons of water
3 lb. hops ⎭
(a pinch of black malt adds colour to the beer)

The water was put into the mash-tub after boiling it in the copper, seven pailfuls of boiling water to three pailfuls of cold. The malt was then shot on to the water and carefully stirred, then the tub was filled with water. After leaving for four hours the resultant liquid "sweet wert" was drawn off and put back into the copper with the hops to be boiled for four hours. This was strained into a second mash-tub when the beer was lukewarm, yeast was added and the beer was thus "set to work". It could now be left to cool and afterwards skimmed and "turned" into the barrel. A "second wert" could be brewed by adding water to the grains after the first brew had been drawn off. Horses and cows relished the grains, and the spent hops became a good fertilizer, nothing being wasted.

I became a good dry-hand milker at an early age; machine-milking was in its infancy and did not reach us. We poured the milk into flat shallow pans (see sketch) and left it until the cream had risen to the top. It was then skimmed and the cream kept for a few days ready for churning. Churns were varied even in those days and I have sketched a box churn of the type my mother used.

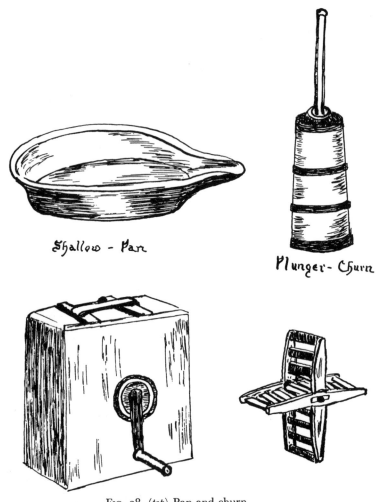

Shallow - Pan

Plunger - Churn

Fig. 38. (*top*) Pan and churn
(*bottom*) Box churn

I also show an older type plunger churn. Later, at Chadacre, I used the end-over-end barrel type and at Byham Hall, Great Maplestead, I have used a barrel churn so suspended as to turn sideways. For some time I milked seven cows there night and morning and churned twice a week. We used a cream separator

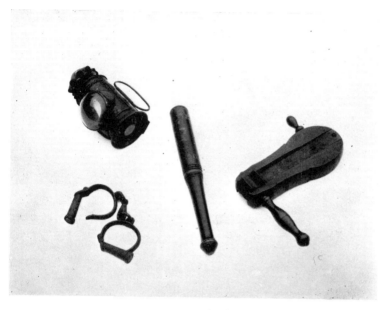

POLICEMAN'S SET

MANTRAP

TRACTION ENGINE MODEL

PORTABLE ENGINE MODEL

FIG. 39. Yoke and wooden bucket

instead of the shallow-pans method. Modern forms are still in use. Churns worked by horse-power were common in the mid nine-teenth century and a fine example from Buckinghamshire is now in the Science Museum. Water-power was sometimes used as for the thresher or the mill, an immense spur-wheel being rotated for the motive power. I also include a photograph showing the butter-making "hands" and a revolving mould which made a nice patterned finish to the butter pats. I heard of an ingenious farmer once who made a barrel-churn with a place in the middle where his small boy sat astride. The machine moving up and down formed a dual purpose—the necessary churning movement and a rocking-horse for the youngster. Early methods for the sale of butter are of interest. In Suffolk it was sold by weight and my photo shows a very old set of wooden scales used in the 'Butter-Market' at Ipswich and Bury St. Edmunds. In Norfolk it was sold by the pint and half-pint, being compressed into the mould. Also on the same picture is the Essex butter-basket formerly owned by the late Mr. Saggers of Radwinter and given to me by Miss Halls, when butter was sold by the yard and half-yard measurement. It is after all not so many years ago that the working dairy was a feature at all agricultural shows. Men and women participated and whereas ex-Chadacre students predominated at the Suffolk Show, ex-Writtle students formed the biggest number of competitors at the Essex Show. Another subject taught at Chadacre during my terms there, 1922–4, was cheese-making, under the capable tuition of Miss Spurr, soon to become Mrs. Brieant when she married the Vice-Principal. Cheese is a very ancient commodity in fact and Job asks, "Hast thou poured me out as milk and curdled me like cheese?", and David was sent by his father Jesse "to carry ten cheeses to the captain of their thousand in camp and to see how his brethren fared". Cheese is also mentioned by Homer, Euripides, Theocritus and other early poets. Possibly the Romans (who were good cheese-makers) introduced the custom into Britain. They simply allowed the milk slowly to sour without the addition of rennet. It would be beyond the scope of this book to describe the varied methods of making the great assortment of British farm cheeses. Among the implements needed was a cheese press for forcing out the whey and helping to solidify the cheese. The early variety was a long timber lever so adjusted as to impose the weight direct on to a cheese vat. The end of the lever was sometimes fixed in a hole in the dairy wall, and the weight was merely some large stones. Another type had a large square stone suspended by a screw between the side-posts of a timber frame, the cheese vat being placed directly beneath the stone. The turning of the screw induced gradual pressure. Variations of this type of press are in use wherever farmhouse cheeses are made today. The vat of course was a miniature tub, but sometimes it was made of tin. I illustrate an old cheese cutter with two handles. On many old-time farms butter and cheese-making, together with poultry-keeping and bee-keeping, were sidelines whereby the ladies of the household made their "pin-money".

The book would be incomplete without a reference to shooting. I still use my father's old gun, a 100-year-old Wesley-Richards with barrels inside still like polished aluminium. The game carrier is early Victorian and birds could be carried by the legs only and not uppermost which latter method is said by experts to be correct. The old laws against poaching were very severe and I also picture my man-trap sent me by Mr. G. H. Pockson

of Compton Martin, Bristol. I recall seeing an old painted board
worded "Poachers beware, Mantraps and Spring-guns are set
in this plantation". Speaking of the law reminds me to record a
little group of objects not strictly agricultural which find their
place in my collection. There is a policeman's set comprising a
truncheon on which is painted the Royal Arms and the words
WILLIAM IV, 1830. There is his bulls-eye lantern, his rattle (fore-
runner of the police-whistle) and handcuffs. Mr. C. F. Turnbull
also gave me a Bow-Street runners' lamp, which has been to
Scotland Yard for identification but which is not quite complete.
There is also a fine fireman's helmet and leather fire-bucket, as
well as a large magnificent model of a Shand-Mason steam fire
engine made by the late H. S. Goodman of Woodford, a traction
engine and a portable engine by the same master hand. There are
some model windmills made for me by Charles Rawlins of Herne
Bay from my own sketches and detailed measurements, and an
oil lamp used by Burrell's boiler-makers, in their steam engine
works. A windmill template and damsel from the late Maurice
Holgate, a roasting-jack, and a penny-farthing bicycle just about
complete the collection. In my last Chapter, I want to record
words and sayings which I have collected during the past thirty-
nine years, all of which are linked with the farm and with country
life.

CHAPTER TEN

In my profession I find I am never divorced from the soil for any long period. Even on Sundays, when I follow the Second Lesson by the reading of the Banns, I am at once linked with the terms used by the farming fraternity of Anglo-Saxon days. "Bachelor of this parish" at one time meant a farm-labourer or cowherd. "Spinster" reminds us that in the days of King Alfred no woman was considered fit to be a wife until she had proved herself to be a spinster, i.e. had spun herself sufficient thread with which to weave a set of body linen, bed linen and table linen. A lot of farm history is in fact preserved in our everyday conversation. An acre was in the past any enclosed field, and to this day many of us term the churchyard "God's acre". The words "Lord" and "Lady" originally meant "loaf guardian" and "loaf-kneader", the obvious first duties of man and wife. Barley meant "bread-plant" and "barn" was bread-house in earlier days. After the Norman Conquest a peculiar transition took place in that the Normans used their own words for food, whilst their Anglo-Saxon slaves kept their old terms for the stock which they tended and which provided such food. Thus the Anglo-Saxon "ox" became the French "boeuf" (beef). Sheep became "mouton" (mutton), calf became "veau" (veal) and pigs or swine became "porc" (pork).

I formed an early habit of recording words, dialect or sayings of my boyhood and this chapter is the result. All the words and sayings will be those I have actually heard or even used. The farm kitchen or scullery was called "backhouse", pronounced "back-hus", and in the backhus at Rivetts Hall, Hartest, hung an ancient "hake" over an open fireplace. I have a specimen which Harry Cranwell gave me and which is in reality a stew-pot hook. Outside the "backhus" at Whepstead Hall came milk customers with their tins to purchase a pennyworth of separated milk. The boy who chopped sticks and brought water and coal was known as "back-hus boy". My mother's apron was termed a "mantle" and my father's leggings were known as "buskins". A floorcloth was a

AUTHOR JUDGING CART-HORSES AT THE REGENT'S PARK PARADE, 1959

DRAINAGE SPADE, WOODEN FENLAND SPADE AND DOCKING IRON

"dwile" and farm boots were called "high-lows". A knitted jersey was a "ganzy". The act of skimming the cream from the milk was known as "fleeting the milk". "Together" merely meant all of you, a boil on the back of the neck was a "push". I cannot think of a more expressive word either. A lazy man was said to have "soft shoulders", to "sag" meant to decline in health, and I have been called "cack-handed" for holding a pitchfork in an unorthodox manner. A man who sold clothing from door to door was called a "pack-man" and travelling gypsies were "diddie kie". When an animal died on the farm we sent for the "knackers-cart".

A heavy dew was called a tidy "daig"; a cold foggy morning was said to be "rafty", "tetchery" weather was wet and dull; the heavy storms which often occur in the late winter and early spring were termed "lamb-storms". Paving-stones in the backyard were either "flags" or "pamments" and the little cabin down the garden was either the "privy" or "petty" and the department behind it was a "bumbay". I ought to know this for I fell waist deep into one at Whepstead Hall and had to be rescued. Bees were kept in a straw "skep", the small shrew mouse who tried to get at the honey was a "Renny". The plural of mouse was "meese", the huge beetles who fly by night were "midsummerdoors", and a dragonfly was a "hoss needle" because no doubt of the needling effect on a horse if they happened to sting him!

I soon learned that the heavy scraper in which I cleared up the farmyard mud was a "slud-hoe", and if a person was thought to lack sense he was said to have been "hit with a slud-hoe". A clover stack was always a "stover-stack", a newly weaned calf was a "wennell" and the smallest pig in a litter was "cadman". If a field were entered by a bridge-ditch it was called a "whellum", a drain was a "grip" and to dam it up was to "stank" it.

A tiny load of corn or hay was a "buck" load or "jag", and when loading it I would be told to "jounce" on it to pack it down. We had a spirited trace-horse called Blossom who was given to pulling all out in her collar until she sank from exhaustion. I was always urged to check her in case she "swounded"—a corruption of swoon surely?

If a field was easy to get at, it was said to "lie gain", peastraw was "pea reis", a wood was often known as a "link" and a common was a "tye". This word figures in many place names to-day. "Four-a-leet" denoted cross-roads, and "stingy" meant to be mean. "Battlings" was a term used to describe a quantity of brushwood too small for timber and too big for faggots. If rotten, these pieces were called "sear" and the low undergrowth of a wood was the "slop", and a "drift" was a lane between fields. A "brief" was a special collection announced by the Parson for someone who had suffered calamity, fire, theft, or bereavement for instance. To "lie by the wall" meant to be dead but not buried. Some sayings are obviously of French origin; for instance, an old lady would say to me, "I aren't avised of it", when she meant she wasn't sure. To stare at a person was to "garpe" (corruption of gape), and a lively child was said to be a "limb for mischief". I recall spilling some water when staying with an aunt who said, "that don't signify", which really meant, "All right, it doesn't matter." There were several ways of describing something large; it could be "good tightly big", "wholly mortal big" or "nation big".

A silly person was called "sawny", "lary", or just a "tulip". To "crake" was to boast, a noisy child would be told to "shut up as he was enough to dunt the devil". To get a taste of anything was to get a "say". Yonder was always "hinder" and if a person wanted to prophesy a hard winter he would say, "we're in election for a hard winter." To hurl a stone was to "ding" it or to "hull" it and to be frozen was to be "frawn". The plural of house was "housen". "Sales" meant seasons and we still speak of "wheatsel" and "hay-sel". The threshing machine was the "chine" and in those days we went "fraishing for nawthen". "Enow" was enough, a see-saw was a "teeter-me-tawter". The use of the word "do" was peculiar —"Hurry up do we shall be late" literally meant, "Hurry up otherwise we shall be late". Likewise "I fare to fancy" meant "I rather think" and seems to come from the German "fahren". Wedding guests were "weddeners" and chapel folk were "meeteners". It may well be that some of these words were peculiar to rural East Anglia, although many I am sure belong equally to other districts. Those of us like myself fifty-two years old and over, find ourselves lapsing into the vernacular from time to time. Not long ago I met a man in London who said he knew I was from Suffolk because I had just told him "that's going to rain".

Bee-keepers too had quaint sayings and customs, and if a member of the household died they would go and tell the bees and put a piece of black crêpe over the hive to prevent another death.

This is all I have to say. I have tried to write of the people we are proud to remember as our forefathers, the tools they used, the work they did, the language they spoke and the animals whose co-operation they obtained.

We owe them an enormous debt, for they have made us what we are. They believed that farming was more than mere work, and saw in it a vocation and a way of life, believing in a very simple philosophy "Live as though you would die tonight, farm as though you would live for ever."

INDEX

94